A History of the Second Fifty Years
American Mathematical Society
1939-1988

AMERICAN MATHEMATICAL SOCIETY
CENTENNIAL PUBLICATIONS
Volume I

A History of the Second Fifty Years
American Mathematical Society
1939-1988

by
Everett Pitcher

American Mathematical Society
Providence, Rhode Island
1988

Library of Congress Cataloging-in-Publication Data

Pitcher, Everett, 1912-
  A history of the second fifty years, American Mathematical Society, 1939–1988.
  Includes index.
  1. American Mathematical Society—History.
  I. American Mathematical Society. II Title.
QA1.P52 1988   510′.6′073   88-22318

ISBN 0-8218-0125-2

CONTENTS

Preface

The history of the first fifty years of the American Mathematical Society was prepared by Raymond Clare Archibald and was published by the Society at the time of the Semicentennial in 1938. The present volume covers the second fifty years. It is not an attempt to write what Professor Archibald would have written had he been here to do so, something that I could not expect to achieve. It is an effort to cover principal functions of the Society and the development of the organizational structure by which it carries out those functions.

The two major activities of the Society are publication and meetings. Following an introductory chapter on the state of affairs fifty years ago there will be several chapters on publications followed by several on meetings. They are not easily separated because the scientific substance of meetings and particularly of conferences is reported through publications. Then there will be several chapters on the organization of the Society itself. However, much of this will have been reported already since the persons who edit publications and handle meetings are members of that organization. Next there are chapters on the functioning of the Society through various committees, though committees will have appeared throughout. Finally, the physical structures and the archives will be described briefly.

Records on which this book is based are described in the first chapter. Such records as minutes of the Council and Trustees and announcements in the *Bulletin* and *Notices* have been paraphrased or quoted liberally without attribution.

A number of bits of biographical information have been secured from *American Men and Women of Science* (or its predecessors) edited by the Jacques Cattell Press and published by R. R. Bowker and Company.

Portions of the book have been written by others, specifically sections about *Mathematical Reviews* by Ralph P. Boas, its second executive editor, and by William B. Woolf, its current and long time managing editor. The tables in the chapter on finances were prepared in the Fiscal Department in Providence. In addition many individuals have provided bits and pieces of information. These include Nathan Reingold, then with the Smithsonian Institution, Andrew Mattel Gleason, past president of the Society, John Willie Green, former secretary of the Society, Lincoln K. Durst, former deputy executive director, and current staff members of the Society. The record is not complete on who the latter are, particularly as a question asked of one is answered by another in a general spirit of helpfulness. However the list certainly includes Robert C. Blanchette, Gary G. Brownell, Tricia Cross, Monica Foulkes, Regina M. Girouard, Ellen Heiser, Edith Krekorian, William Judson LeVeque, Muriel Scribean, and Ben Silver in Providence and Robert G. Bartle and Jane E. Kister in Ann Arbor.

The editorial management and support was provided by Mary C. Lane in Providence.

I thank all the persons named and unnamed who have helped me.

One who should not go unnamed but should be placed in a separate paragraph is my secretary Ruth Hahn, who has lived through the trials of word processing in preparation of the manuscript and deserves special thanks.

Introduction

RECAPITULATION

This is a history of the second fifty years of the American Mathematical Society. It is convenient to summarize the first fifty years and in particular the state of affairs in 1938.

The purpose of the American Mathematical Society is stated in its certificate of incorporation, dated 3 May 1923, "into a corporation pursuant to, and in conformity with, the provisions of sub-chapter III of 'An Act to Establish a Code of Law for the District of Columbia' approved March 3, 1901, and of the several Acts amendatory thereof." The statement is that "The particular business and objects of the Society are the furtherance of the interests of mathematical scholarship and research."

The Society accomplishes its purposes through meetings and publications and it is to these two topics that a great deal of this history is devoted.

An effort will be made to give a picture of the Society of 1938 and an indication of the political and social influences that shaped its subsequent development. In another direction, the organizational development of the Society, the financial growth, and the changing physical circumstances will be treated. Significant personalities in these developments, notably the presidents, will be described briefly.

The American Mathematical Society dates its founding from 1888. After some preliminary planning, the organization meeting of six persons was held on 24 November 1888 (Thanksgiving Day) at Columbia College. The constitution was adopted at the second meeting on 29 December 1888. The organization was initially named the New York Mathematical Society, for the London Mathematical Society was a predecessor and a model. The name was changed to the American Mathematical Society in 1894 by an amendment to the constitution.

Initially, the Society was small but grew rapidly. There were 16 members by the end of 1889. The numbers in some subsequent years were:

1890	230
1900	3470
1910	6300
1920	770
1930	1926.

The year 1923 saw a reorganization of the Society as a corporation under the laws of the District of Columbia. There were then 1250 members. By 1938, the year of the Semicentennial Celebration at Columbia University, the membership had increased to 2139.

The celebration took place in the interval 6–9 September 1938. It was the forty-fourth Summer Meeting, with an attendance of about 700, including 419 registered members of the Society. The public record is presented in several parts. The conventional report of the scientific meeting by Associate Secretary T. R. Hollcroft is published in the *Bulletin of the American Mathematical Society*, 44(1938), 744–755. The account of the celebration was prepared by the secretary and the associate secretaries. It is in the *Bulletin of the American Mathematical Society*, 45(1939) and consists of a frontispiece in three parts, namely a photograph of those assembled, and a report on pp. 1–9. There are twelve appendices on pp. 10–30, the first being a letter from President Franklin D. Roosevelt, and a brief history of the Society by Raymond Clare Archibald on pp. 31–46. Archibald was the chairman of the subcommittee on the program and the brief history is the published version of his address.

There was one unofficial and previously unrecorded event at the semicentennial. The celebration featured a gala dinner on Wednesday, 7 September 1938, at the Hotel Astor, attended by 368 persons. The charge for the dinner was $3.00. This figure should be compared with the fact that single rooms were available to registrants for $1.00 and $1.25 and the Astor itself charged $3.00. There were younger members of the Society who thought $3.00 outrageously large for a dinner. Accordingly there was also an impromptu event known as the "ungala dinner" in a restaurant on 42nd Street, attended by at least twenty-five mathematicians, who sent a telegram of congratulation to the gala dinner. At least two future winners of Steele Prizes for the cumulative effect of their lifetime of mathematics, one future president of the Society, and other future officers were among the participants.

The Semicentennial was commemorated in the *Semicentennial Publications* in two volumes. The first, cited as [A], is a history written by Professor Archibald and consists of ix+262 pages. The second is a collection of nine addresses, one by George D. Birkhoff on Fifty Years of American Mathematics and the other eight on broad fields of mathematics. It consists of vi+315 pages. In the Council minutes of 28 December 1938 is the final report prepared by Thornton C. Fry and concerned principally with the finances of the venture. Net expenses of the celebration over and above those attributable to a usual summer meeting were $343.29. The celebration was subsidized by Columbia University to the extent that it rendered no bills for services. The publications came in below budget at $4,645.25. The report is necessarily silent on the amount recovered through subsequent sale of the two publications.

The total annual disbursements of the Society did not pass $10,000 until 1923 and were somewhat more than $30,000 in 1938.

There will be a number of references to dollar amounts in this account. These must be interpreted in the light of inflation over a period from 1938 to 1988. There are well established measures of the general inflation. These may not exactly measure the effect of inflation on the goods and services supplied to the Society and by the Society. Roughly speaking, the inflation between 1938 and 1988 is a factor of 8.4 as measured by the Consumer Price Index.

The elected officers in 1938 consisted of a president, three vice presidents, a secretary, four associate secretaries, a treasurer, a librarian, twelve members of the three editorial committees, and fifteen elected Council members. Certain past officers remained as Council members after their terms as elected officers expired. There was a board of five elected Trustees.

The only meetings of the Society in the late 1930s were general meetings. Two of these were national in character, a Summer Meeting in early September and the Annual Meeting between Christmas and New Year's Day. These two were joint meetings with the Mathematical Association of America. Other meetings, of which there were sometimes seven, were regional, there being three areas known as Eastern, Western, and Far Western.

The usual fare consisted of an invited address and contributed ten minute papers. Occasionally there was more than one invited address, sometimes two or three addresses in a single field. The number of contributed papers was sometimes as small as 10 and exceeded 50 only at the two national meetings. The Colloquium Lectures were a feature of the Summer Meeting. The retiring presidential address, required by the bylaws, usually came at an Annual Meeting. The Annual Meetings were sometimes held in conjunction with the annual meeting of the American Association for the Advancement of Science and such events of Section A of the Association as the retiring address of the section chairman were scheduled in connection with the Society meeting.

The staff of 1938 consisted of four persons, who had office space in Low Memorial Library of Columbia University, where the library of the Society was also housed.

The publication program of 1938 was quite simple. The *Bulletin*, which was started in 1894, published volume 44 with 888 pages in that year. It was a privilege of membership. The journal then carried reports of meetings, abstracts of contributed papers, lists of new publications, reports of meetings, obituaries, book reviews, personal notes, and lists of doctorates (72 in 1937), as well as invited and contributed research and expository papers. It served functions now covered by the *Bulletin* (New Series), *Abstracts, Notices*, and *Proceedings*. The *Proceedings* was foreshadowed in the reorganization of the *Bulletin* that took place in 1931, with the separation within one journal into alternate "gray issues" of mathematical papers and "green issues" of news, addresses, and abstracts.

The *Transactions*, first published in 1900, produced volumes 43 and 44 in 1938, with a total of 1073 pages. It was a subscription journal.

These two journals constituted the total of the periodical publication program of the Society. The *Duke Mathematical Journal*, published by Duke University beginning in 1935, had received considerable encouragement from the Society and from its beginning and for many years thereafter had two editors named by the Council.

Beginning in 1927, the Society provided some editors for the *Annals of Mathematics*, published by Princeton University, an arrangement that persisted until 1965.

The *American Journal of Mathematics* began publication in 1878 before the New York Mathematical Society was founded. Early in the history of the Society, a cooperative venture with the *American Journal* was sought and failed. A second attempt resulted in an agreement effective in 1927 with volume 49, whereby the Society shared in the financial management with the John Hopkins University Press and supplied some of the editors. The cooperation in publication was terminated effective in 1977 but not the arrangement to supply editors.

The first book published by the Society was the *Proceedings of the International Congress* of 1893 in Chicago at the time of the Chicago World's Fair. The Colloquium Publications first appeared in 1905. The year 1938 saw the appearance of volume 22, *Summable Series and Convergence Factors*, by Charles Napoleon Moore. Up to 1938, no other books had been published by the Society.

Total receipts in the interval 1891–1937 were $578,366.16 and total expenditures were $534,109.17. This compares with the fact that the annual budget passed ten million dollars in 1985. Dues over the 37 year period accounted for $237,552.52, i.e. about 41% of income. This differs greatly from more recent times, when dues account for about 9% of income. Cost of publication in the 37 year period shows up as $401,798.45, i.e., about 75%

of expenditure. This agrees closely with the current percentage, which was 73% in a recent year.

The change in the percentage of income from dues suggests that recently a small percentage surplus on a large publication program now pays for a variety of services once covered if at all by dues.

ORGANIZATIONAL RECORDS

Publications

The scientific records of the Society consist of the journals and books that it has published. These will be discussed at another point. The organizational records are of several kinds.

Beginning with volume 1 (1892), the *Bulletin of the American Mathematical Society* records a summary of each meeting, including the names of speakers with the titles of their papers. Moreover, there are lists of bibliographic information on papers read and subsequently published. With the passage of time, these reports came to include summaries of papers delivered. (The predecessor *Bulletin of the New York Mathematical Society* did not include such material.)

There were advance announcements of meetings but the record of them is no longer complete. There are four bound volumes of these. (In 1987 they were in the possession of the Secretary as part of the material scheduled for the archives in the Brown University Library). These begin with the announcement of the Eighth Summer Meeting and the Third Colloquium in Ithaca in August 1901. Next comes the Thirteenth Summer Meeting and Fifth Colloquium in New Haven in 1906. Beginning with the 172nd Regular Meeting in New York on October 31, 1914, there are many announcements and programs but the set has not been examined for completeness. The first book ends in 1926. Subsequent books cover 1927–1938, 1939–1947, and 1948–1953. The later volumes give the appearance of being quite complete. The function of announcing meetings and listing programs was taken over in 1954 by a new journal, the *Notices*.

The reports of meetings contained items called abstracts, written in the third person as though by the secretary preparing the report of the meeting. In the case of a paper already published by the time the report of the meeting appears, there is merely reference to the published paper. The format changed with volume 27 (1920–21) to that of abstracts prepared by the author. Beginning with volumes 36, author abstracts were published in advance of meetings (after a transitional period to clean up abstracts of papers already presented). The last of these in the *Bulletin* appeared in volume 63 (1957).

With volume 5 (1958) of the *Notices*, author abstracts were published here instead of in the *Bulletin*. This state of affairs persisted through 1979. In

1980, a new journal, *Abstracts*, came into being. Initially it was distributed on request to members without charge but with volume 2 (1981) it became a subscription journal.

The reports of business meetings of the Society were in the *Bulletin* from the beginning through volume 83 (1977). The *Bulletin* was the journal of record and was a privilege of membership. The *Notices* was also a privilege of membership from its inception and, by action of the Council, became the journal of record effective with volume 25 (1978). From that point on, it contained the reports of the meetings.

Minutes of the Society and of the Council contained appointments to committees and appointments of representatives, such as delegates to college and university inaugurations and anniversaries. These were published in the account of meetings as well, through volume 78 (1972) of the *Bulletin*. Publication of lists of representatives in Council minutes ceased at this point.

The Society began a publication called the *Directory of Officers and Committees* in 1954. In 1957 the title was changed to *Administrative Directory*. The publication included the certificate of incorporation, the bylaws, the past winners of prizes, the special funds of the Society, and the lists of past presidents, Gibbs Lecturers, and visiting lecturers. The lists were dropped after 1960. The publication appeared in 1961 as published for the National Register of Scientific and Technical Personnel and included officers and committees of 14 organizations in the mathematical sciences, together with a list of department heads at more than 500 colleges and universities. The year 1965 was the last year of National Register sponsorship, after which it became again a Society venture, with modest financial assistance from some other organizations. The amount of information increased with time.

There was a brief period after 1972 during which committee appointments were not published except in the *Administrative Directory*. By 1975, they were being listed as news in the *Notices*, a feature that became formalized with time, so that they are an entry in the annual index of the *Notices* beginning in 1977.

Members elected were noted in the minutes of the Society and the minutes of the Council from the beginning and lists were published in the *Bulletin*. There is a bound volume containing lists of members for the year 1898, 1902, 1904, 1905, 1906, 1907. (In the possession of the Secretary in 1987). These lists are supplemented with the constitution, the bylaws, and the reports of the treasurer, the auditors, and the librarian. The volume for 1902 is called *Annual Register*, as are subsequent volumes. The list for 1907 is also bound with volume 13 of the *Bulletin*, that for 1909 with volume 15, and for 1911 with volume 17 (in the bound set in the possession of the Secretary in 1987). Each of these is called an *Annual Register*. *The Annual Register* for other years seems to be unavailable.

There is a List of Officers and Members for 1919–1920 bound with volume 28 of the *Bulletin*. Whether it is the first in this format is not known. The list for 1931–1932 is part 2 of an issue of the *Bulletin*, an arrangement that continued through 1950.

Beginning in 1952, there is a separate annual publication called the *Combined Membership List*. It covers AMS and MAA until 1955, when SIAM was included as well. In the issue of 1971–72, the Canadian Mathematical Congress was included for a single year. It reappeared in 1988.

Minutes

There are minutes of the American Mathematical Society covering the period 1888–1912, bound in two volumes. Pages 1–89 of the first volume and unnumbered pages are minutes of the New York Mathematical Society up to the point in 1894 that the name was changed. Pages 90–281 and unnumbered pages of the first volume are minutes of the American Mathematical Society through 1901. The second volume, with pages numbered 1–231 and unnumbered pages cover the period 1901–1912. Here and throughout, it is the pages of actual minutes that are numbered while attachments are unnumbered. (There are occasional exceptions.) The minutes of the Council of February 24, 1917 note that there was a fire in October 1914, though there is no current record of it. There is a pencilled note that there were no minutes of the Society after 1912. The minutes of the Society of 1901–1912 show fire damage.

The function of minutes of the Society, as already noted, was taken over by the reports of meetings in the *Bulletin* and, beginning in 1977, in the *Notices*.

The minutes of the Council begin in 1891. The first volume consists of pages numbered 1–106 and unnumbered pages covering the period 1891–1899 and pages numbered 1–155 and unnumbered pages covering the period 1899–1907. The minutes of the Council of February 14, 1917 state that minutes of 1907–1912 were lost in the fire. The next available volume covers the period 1914–1923 and consists of pages numbered 1–197 and unnumbered pages. It concludes with two pages recording the business meeting at which the transfer of property from the Society to the incorporated Society was authorized. The minutes of the Council continue with four volumes, with pages numbered 1–616 and unnumbered pages, covering the period 1923–1937. The next set of volumes consists of numbered pages 1–939 and unnumbered pages and runs from 1937 to 1966. Beyond this point, minutes are kept in sets of two years, separately numbered and corresponding to terms of the secretary.

With the passage of time, the relative bulk of attachments has become larger. With the gradual shift of the supervision of details of operation from the Council to the Executive Committee and Board of Trustees, the joint minutes of those two bodies have become very large.

The Trustees had separate minutes, beginning in 1923, when the Society was incorporated and the Trustees came into being. There is a set covering the period from October 1923 to January 1938 with pages numbered 1–125 and unnumbered pages of attachments. There are two subsequent sets. Set A (by inference the secretary's copy) covers the interval December 28, 1937–November 12, 1958. Set B (presumably the treasurer's) covers January 25, 1938–December 11, 1966. The two are not quite identical, particularly with respect to attachments. Set B was augmented by the addition of identified photocopies of the material in set A but not set B. From 1967 on, the minutes of the open sessions of the Trustees are all contained in attachments to Council minutes.

There are minutes of executive sessions of the Executive Committee, of the Executive Committee and the Trustees jointly, and of the Trustees alone. There is the possibility that they are not complete. In particular, there may be minutes concerning salaries that were filed only with the fiscal manager in support of disbursements.

Files

There are two file cabinets of early correspondence. There was an incomplete attempt at rearrangement, so that its present state (1987) is as follows:

One drawer of loose paper.

Five drawers of alphabetic folders of 1930's and 1940's, these being secretarial files of R.G.D. Richardson and John Robert Kline.

Two drawers of material classified by subject or author and date. The blocks of time are 1920–1930, 1931–1940, 1941–1945, 1946–1950, 1951–1955. There are separate files on Annual Register, 1892–1904, and on individual International Congresses through 1954.

The files of Secretary E. G. Begle, 1951–1956, and of Secretary J. W. Green, 1957–1966, are in drawers by year as are those of Secretary Arthur Everett Pitcher, 1967–1988. As the bulk of files grew, there are separate files by meeting for the Executive Committee and Board of Trustees.

The introduction to [A] refers to fifteen years of the files of Thomas Scott Fiske. The writer has not found either these or other files prior to the days of Richardson as secretary.

Presidents have kept their own files and some at least of these have disappeared as the president changed. For the most part, copies of correspondence of presidents appear in the files of the secretary but it is clear that not all significant material is preserved in this manner.

The location of files of treasurers is unknown. A major portion of the significant work of the treasurer appears in minutes of the Trustees.

The chapter on Archives is concerned with the disposition of these records.

POLITICAL INFLUENCES

The growth and the changing activity of the Society must be viewed in the light of world history. There were several related influences that profoundly affected American mathematics and the Society.

Recall that Hitler came into power in 1933 and that racial superiority and the accompanying racism were prime tenets. Germany seized Austria and most of Czechoslovakia in 1938. The attack by Germany on Poland came on 1 September 1939 and Great Britain and France declared war on Germany on 6 September. War on a smaller scale had been going on in several places for some time, with the Italo-Ethiopian war in 1935–1936, the Spanish Civil War in 1936–1939, and the Sino-Japanese War in 1937–1941. Mussolini governed Italy from 1922. He initially distrusted Germany under Hitler and espoused neither socialism nor racism. His position changed and the "Rome-Berlin Axis" was established. By 1938, Italy had accepted racist laws. In 1940, Italy entered the war on the side of Germany and France capitulated.

These political events disrupted the lives of mathematicians and the development of mathematics. One effect was a substantial displacement of mathematicians, including emigration to the United States. Another was the beginning of large-scale government financing of research and of education in the United States.

The depression of the early 1930s brought about unemployment in academic ranks. Those employed found their salaries fixed, or, in some cases, reduced. At the same time, the political unrest in Europe, of which anti-Semitism was a substantial factor, caused immigration of persons with the means or the talents that permitted it. There was a substantial number of mathematicians who entered the United States prior to World War II. Reingold has placed the number between 120 and 150 in the interval 1933–1941. This number should be compared with the annual U.S. Ph.D. output in mathematics of about 75. These immigrants tended to be well educated competent mathematicians. There were several effects. The strengths of some of the immigrants lay in their research. Their presence initially improved the quality of mathematics where they went but denied immediate opportunities to home grown talent. They were sought or tolerated on scientific or humanitarian grounds and resented by some in economic or chauvinistic terms. Part of the resentment was anti-Semitism. The long-run effect of their arrival was an order of magnitude of development and expansion of mathematical research in the United States. The reader is referred to the paper "Refugee

Mathematicians in the United States of America, 1933–1941: Reception and Reaction" by Nathan Reingold, *Annals of Science*, 38(1981), 313–338.

The federal government endeavored to relieve some of the unemployment of the depression of the thirties through made work. Although some of it was characterized as leaf raking, there were also projects of permanent value. The WPA (Works Project Administration) was commissioned in 1938 "to employ needy professional and technical persons in planning, computing and making available a series of mathematical tables, with the assistance of qualified men and organizations." The tabulated functions included logarithm, exponential, and trigonometric and Bessel functions. The work was done according to procedures approved by the Director and Staff of the National Bureau of Standards. The Director was Lyman J. Briggs and the Technical Director of the Project was Arnold N. Lowan. Not only did the project produce valuable tables but also it was a source of development of numerical analysts among its supervisors.

Although the United States had been preparing for war for some time, its entry was precipitated by the Japanese attack on Pearl Harbor on 7 December 1941. A great deal of work in the application of mathematics during World War II was done directly in government installations by civilian employees and military personnel. Such were the workings of the draft that distinguished mathematicians (e.g. Herbert Federer, Gerhard Hochschild) applied their mathematics in the non-commissioned ranks. The Ballistics Research Laboratory at Aberdeen Proving Ground, where there were 27 Ph.D. mathematicians, was an excellent example of this sort of establishment.

The NDRC (National Defense Research Committee) gathered a group of scientists including mathematicians at Columbia University. Their work covered a variety of applied mathematics, including rocket theory.

Many individual universities, notably New York University, did research in applications of mathematics under contract.

The GI Bill (Servicemen's Readjustment Act of 1944, Title II, Sec. 400 and subsequent amendments) provided educational benefits for returning veterans of World War II. A great deal of the education was undergraduate or vocational but there was a substantial component of graduate education and even some postdoctoral work supported through this channel. Most of the effect of this was over by 1950.

The launching of the Russian satellite Sputnik in October 1957 had a substantial shock effect in the United States, inducing a feeling that the country might be falling behind in significant areas of science and technology and in the attendant dominance of reputation through publicity about achievement. The result was a substantial increase in government funding of research and education and an increased popularity of scientific and technical vocations. The NDEA (National Defense Education Act) of 1958, amended in 1964, was one of the reactions. It provided for a relatively large number of graduate fellowships, ordinarily held for three years. Fellowships were tied to institutions, which received a subsidy accompanying each fellowship. The result was that the fellowships were widely spread among good institutions rather than concentrated at the most prestigious ones, so that institutions were developed as well as individuals.

The Office of Naval Research was a prominent vehicle in the support of academic research in mathematics immediately after World War II. It carried the burden until the establishment of the National Science Foundation in 1950. Other services, the Army and the Air Force, sponsored contracts for mathematical research until their activity was somewhat curtailed by the Mansfield Amendment (PL 91-441, Armed Forces Military Procurement Act of 1971, Title II, Sec. 204, and successor acts), which reads:

> None of the funds authorized to be appropriated to the Department of Defense under this or any other act may be used to finance any research project or study unless such project or study has, in the opinion of the Secretary of Defense, a potential relationship to a military function or operation.

This wording was carried forward from year to year in successor legislation for a time.

The increase in college population in the late 1960s created the opportunity for the support of additional graduate students as teaching assistants and thus yielded additional Ph.D. mathematicians.

The increase in Ph.D.'s brought about by the NDEA and the increase in teaching assistants and the employment opportunities attributable to the increased undergraduate population overshot its mark, especially as the pressure of undergraduate population was relieved. The graduate population decreased, again overshooting the mark. The shortage was exacerbated by the rising importance of computers, which led students with a generally mathematical bent to study computer science with an eye on future employment rather than mathematics. The available capacity in graduate schools in mathematics and the new shortage in teaching assistants brought many foreign students to American universities.

Primary Journals

The publication program of the Society is its largest venture, accounting, as has been noted, for three-quarters of the budget. The primary journals in the program may be viewed in three categories. The most significant, measured in Society expenditure and effort, consists of five journals edited and published by the Society. Less prominent in the budget are three journals with which the Society has cooperated in editorial matters and a handful of journals that the Society has subsidized.

The *Bulletin, Transactions* and *Memoirs,*
and *The Proceedings*

The development of the *Bulletin* and of the *Transactions* should be discussed together, for there were problems that affected both. The *Bulletin* and the *Proceedings* are discussed together because the latter grew out of the former.

The *Bulletin of the New York Mathematical Society* was the first journal of the Society. It began publication in 1892. The initial character of the journal was defined in its subtitle *A Historical and Critical Review of Mathematical Science*, which was maintained through 1930. The name was changed to the *Bulletin of the American Mathematical Society* in 1894 after three volumes and the numbering was started again at one. The history of the earlier years is presented in detail in [A].

The volume 44 of 1938 was dedicated to its long-time editor Earle Raymond Hedrick. It contained reports of meetings, book reviews, notes on conferences, appointments, and deaths, abstracts of contributed papers, and research papers. There was a list of contributed papers (i.e. abstracts) with bibliographic information on subsequent publication, and an index. Finally, the list of officers and members, the bylaws and the report of the treasurer, were a part of the volume in a separate issue.

In 1930, the journal had been separated into gray issues, consisting exclusively of short papers, and green issues, consisting of everything else, though the color differentiation of the cover did not appear until 1931. The supplement constituting the membership list was covered in a bright yellow.

With the appearance of *Mathematical Reviews,* the annual list of papers that had been a feature of the *Bulletin* was dropped.

Publication of the *Transactions* began with volume 1 in 1900 and continued at the rate of one volume per year (except two in 1922) through 1936. There were two per year through 1948.

After World War II, the volume of material offered for publication increased and so did the backlog of the two journals. The Society had a standing Committee on the Role of the Society in Mathematical Publication that considered the problem and expressed the opinion at the Council of 28 December 1948 that drastic action was needed. The initial recommendations were a limit on the length of papers in the *Transactions* and a limit on the backlog of both journals. Both of these policies were to be called to the attention of the journals subsidized by the Society. The possibility of publishing brief summary articles was considered. The new executive director was requested to seek sources of publication funds that did not involve increased dues or subscription prices and to explore the possibility of new journals to be published by other groups. All of this met with Council approval except the limitation of the backlog. However, Gordon Thomas Whyburn, speaking for the Editorial Committee of the *Transactions,* stated that the committee thought that a limit on length of papers would not accomplish the desired result and would have other adverse reactions. "Not only would we be prevented from publishing papers of high merit which exceeded the length limit, but also we would be effectively robbed of the power to insist on the compression of those manuscripts which fall below the limit. Hence we propose, as an alternative, to adopt a strenuous policy of compression on all manuscripts of whatever length submitted to the *Transactions.*" The Council approved such a policy.

An Emergency Publication Committee consisting of Alan W. Tucker, chairman, R. P. Boas, Samuel Eilenberg, J. R. Kline, Rudolph E. Langer, and C. J. Rees was appointed. Tucker presented the report of the committee, previously approved by the Committee on the Role of the Society in Publication, at the Council of 30 April 1949. With some modification the report was adopted by the Council and the committee discharged. There were two sets of recommendations.

First, the gray issues of the *Bulletin* were to be published beginning in 1950 as a separate journal known as the *Proceedings of the American Mathematical Society.* The green issues were to continue as the entire content of the *Bulletin.* The mechanics of selecting editorial committees was established.

Second, a new publication called the *Memoirs of the American Mathematical Society* was to be created. The *Memoirs* was to contain long papers or sets of cognate papers. It was to be a nonperiodical serial publication (i.e., a paperback book series) sold as separate volumes. The Editorial Committee of the *Transactions*, enlarged by another member, was to edit the *Memoirs*. The printing of the *Memoirs* was to be done by photo offset from author prepared copy, for which instructions were to be prepared. The *Memoirs* was to be started by transferring good long papers or sets of cognate papers, with the consent of the authors, from the backlog of the *Transactions*, costs of retyping to be borne by the Society. Finally, a grant from the Rockefeller Foundation or other source was to be solicited for start-up costs, though the series was to be self-supporting. The editors of the *Transactions* were asked to set higher standards for longer papers and not to accept papers in excess of 75 pages for the *Transactions* once the *Memoirs* was started. Initially the *Memoirs* was to have a separate editorial committee and the change in the bylaws was drafted to provide for it. However, an ad hoc Committee to Recommend Policies for a Separate Memoirs Editorial Committee, consisting of R. P. Boas, S. C. Kleene, and J. L. Doob, chairman, recommended to the Council of 28 December 1952 against the separation and in favor of close cooperation with the *Transactions*. The editorial supervision of the *Memoirs* was left with the Editorial Committee for the *Transactions*.

In September 1949 the Council voted that the invitation to deliver the Gibbs Lecture no longer carry with it the assurance that the lecture will be published and relieved the *Bulletin* of that obligation. In April 1952 this position was reversed. The invitation to give the lecture was to include an invitation to submit a manuscript for publication without being refereed.

The Council of September 1949 gave the *Bulletin* and *Transactions* editors the authority to give priority to papers of exceptional quality if they chose, rather than publish strictly in order of acceptance.

In the report for 1949, the *Bulletin* editors continued to decry the large backlog. The *Transactions* editors however, stated that the *Memoirs* had relieved its backlog problem but that the start-up costs of the *Memoirs* were larger than expected because of the costs of retyping the initial papers for photo offset.

With the year 1950, the gray issues of the *Bulletin* assumed independent identity as the journal *Proceedings of the American Mathematical Society*. Volume 55 from 1949 is the last volume of the *Bulletin* with the dual character. In 1950 the editors of the *Proceedings* reported that their backlog had been reduced but was still large and requested additional pages for the coming year. By the end of 1951, the *Bulletin*, the *Proceedings*, and the *Transactions* all stated that their backlogs were small.

"Blind Refereeing", which is a procedure in which the referee works from a copy of a manuscript from which the name and institution of the author are deleted, surfaced in the *Proceedings* in 1974 and aroused passionate responses. It is a procedure that had never been used in Society journals. At the Council of 22 January 1975, the issue was raised by Mary W. Gray. On a motion by John W. Milnor, the Council voted a two year trial of blind refereeing to be begun at the discretion of the editors and to be reported to the Council.

There were substantial reactions. One associate editor, Norman Blackburn, objected that he had not been consulted, that blind refereeing was contrary to the terms under which he had agreed to serve, and that he objected to it on principles that he delineated. On the other hand, he did not wish to resign and proposed to ignore the instruction to use blind refereeing.

When Saunders Mac Lane was asked to referee a paper under conditions of blind refereeing, he declined, arguing against blind refereeing at length on principle. Other potential referees declined to serve.

Peter L. Duren resigned as associate editor, stating his reasons.

A motion to abolish blind refereeing failed. A motion to instruct Professor Blackburn to institute blind refereeing was set aside in favor of a motion apologizing to him for having instituted the policy without consulting him and asking him to continue to serve under the ground rules in force when he was selected. That motion failed, leaving the matter unresolved. Professor Blackburn served the remainder of his term.

In April 1977 the Editorial Committee of the *Proceedings* (including associate editors) approved the following statement by a substantial majority:

> Blind refereeing was a completely uncontrolled experiment and there is no way to evaluate it. We would be happy to drop it but not at the expense of dissent within the Society. It is not the editors' responsibility to remove internal evidence in places where it may appear such as the bibliography and introduction.

Some editors were unhappy with blind refereeing and wanted it to be optional. Editors were learning to live with blind refereeing but regarded it as a nuisance. Ronald G. Douglas, an editor, stated that the difference of opinion over its value was not held in the editorial committee with the degree of passion that appeared to rest with some members of the Council.

On a motion by Professor Gray, it was decided to continue blind refereeing. On a motion by Robert G. Bartle, it was agreed to solicit detailed factual experience of authors with blind refereeing in confidence through an announcement in the Notices. There were almost no returns.

In January 1979 on the recommendation of Joseph A. Wolf, speaking for the Editorial Committee of the *Proceedings* as its chairman, blind refereeing as a policy was abolished. At the same time, it was agreed that a paper would receive blind refereeing if the author so requested and provided the additional blind copy.

Not all parties to the controversy were satisfied with this resolution. At the Council of 2 January 1980, Chandler Davis moved to restore the policy of blind refereeing. In the discussion it appeared that requests for blind refereeing had been rare. The motion was defeated. A motion by Murray H. Gerstenhaber to experiment by refereeing some papers twice, once blind, was ruled out of order as not on the written agenda. This appeared to close discussion of blind refereeing.

During the 1960s, for reasons noted in the section on political influences, the amount of published mathematics increased enormously. The government support of fellowships and the increased undergraduate enrollments that in turn supported more teaching assistants led to more Ph.D.s, with publication growing out of dissertations. The *Transactions* was a natural vehicle. There were twelve volumes with a total of 6497 pages in 1969 and the publication continued at almost that pace through the 70s and into the 80s. The *Proceedings* felt some of the burden as well.

The size of the *Transactions* has been governed in part by an attitude of the Council that it is the duty of the Society to provide a place for the members to publish research that meets a reasonable standard and by a policy of attempting to run the journal without a backlog. Efforts in the middle seventies to create a journal of superb quality by very strict standards of acceptance as opposed to a good journal with high but reasonable standards were resisted.

In part the size of the *Transactions* is dependent on its policy with respect to short and very long papers. There is an agreement between the *Proceedings* and the *Transactions* that the former takes only shorter papers and the latter only longer papers. The cut occurred for many years at about eight printed pages but this number has increased toward twelve. Very long papers, as noted, become *Memoirs*.

A periodical has an advantage over a book series in the continuing nature of a subscription. Effective in 1975, the *Memoirs* was converted for that reason to a journal, although individual volumes are also sold as a book series.

From the beginning the *Bulletin* was a privilege of membership although it had also a small paid subscription list among individuals. There was a movement in 1985–1986, spearheaded by Susan J. Friedlander, to amend the bylaws to allow individual dues-paying members to request not to receive the *Bulletin* and to have a reduction in dues to correspond. There was an opinion poll of the membership in 1986 that showed support for such a proposal although there was a segment of members of the Council who thought that an organization devoted to research in mathematics should distribute some of the product to every member. Close examination of costs of production showed that the realistic dues reduction could be no more than five dollars (the marginal cost of a year of the *Bulletin* being $2.93). Upon receipt of this information the Council of 4 August 1987 declined to recommend amendment of the bylaws to allow for a change in privilege of membership.

The character of the *Bulletin* has changed several times. A department called Research Problems began in 1954. The announcement read as follows:

> The department of Research Problems will publish the statements of problems whose solution would make a significant contribution to mathematical research. Problems which are suitable for publication in the problem department of the American Mathematical Monthly will not be accepted for publication in Research Problems. Only problems whose solutions are unknown to the author should be submitted. Furthermore, the problems desired are those for which the solution will take the form of a research paper to be accepted on its merits and published in a research journal; since the *Bulletin* does not accept contributed papers, it will not publish the solution of its research problems. An attempt will be made, however, to publish references to papers which contain solutions.

The department was not represented in every issue. The desirability of the department was questioned. For example, a motion from the Editorial Committee to discontinue it was defeated in December 1956. The last occurrence of the department seems to have been in 1966.

The department of Research Announcements was established in 1958. The heading was the following:

> The purpose of this department is to provide early announcement of significant new results, with some indications of proof. Although ordinarily a research announcement should be a brief summary of a paper to be published in full elsewhere, papers giving complete proofs of results of exceptional interest are also solicited.

The change was timed to coincide with the transfer of abstracts from the *Bulletin* to the *Notices*. Papers were to be submitted directly to members of the Council, who served as referees (or secured referees). The name of the Council member appeared with the notation "Submitted by". As a corollary, Council members had direct access.

With time, the Editorial Committee became dissatisfied with the system. The number of announcements increased and not all met the desideratum of "early announcement of significant results." The Council of 19 April 1968 passed a motion that papers be submitted to the editor in charge of the department, who would use the Council members as a board of referees (a Council member could instead secure a referee). The name of the Council member accepting a paper was to appear with the legend "Submitted by."

This change did not suffice and in fact the system of submitting papers directly to Council members reappeared. In January 1971, the ad hoc Committee to Review Society Activities, consisting of M. Atiyah, Morton L. Curtis, Samuel Eilenberg, Paul R. Halmos, William LeVeque, and Calvin C. Moore, chairman, presented its final report. Among its recommendations was one "[t]hat efforts be made to markedly increase the average quality and to drastically decrease the number of Research Announcements in the *Bulletin* and that firm editorial control be exercised by the editor." The Executive Committee had concurred in strong terms as did the Committee to Monitor Problems in Communication. Key recommendations from these two bodies favored no more than ten or twelve announcements per issue and firm editorial control.

These strictures proved insufficient and the Bulletin Editorial Committee, through a letter from John L. Kelley, made very specific recommendations to the Council of 14 January 1974. As modified slightly by the Council, these were approved by the Council in the following form:

> Research announcements are intended for rapid communication of outstanding results that are to appear elsewhere with complete proofs. It is only exceptionally that a paper containing complete proofs should appear as a research announcement. Generally, if a paper is acceptable to the *Proceedings* or similar journal it should not be accepted as a research announcement.
>
> Each Council member who is a member of one of the Society's Editorial Committees has the privilege of submitting six (6) research announcements per year for publication in the *Bulletin*. Of these six research announcements, at most one may be a complete paper. The editors of the *Bulletin*, in notifying a Council member of receipt of a communication, will inform him, for example, that this is paper number 2 of the six papers he is permitted to communicate during the current year.
>
> Each research announcement is normally limited to five (5) typed pages. More precisely, no research announcement is normally acceptable if it exceeds about 100 typed lines of 65 spaces each.

This was no more successful than previous efforts. At the Council of 11 April 1975, President Lipman Bers requested the Editorial Committee of the *Bulletin* to re-examine the procedure for handling research announcements and to report at the next meeting. At the meeting of 21 January 1976, a proposal from the Editorial Committee to appoint associate editors in fields not well covered was defeated.

The *Bulletin* in its then current numbering terminated with volume 84 (1978), to be followed by *Bulletin, New Series* 1(1979). The plan for the reorganized journal was initiated by P. R. Halmos of the Editorial Committee with the approval of Olga Taussky-Todd and Hans Weinberger, the other two members. He proposed in January 1977 that the *Notices* be the journal of record of the Society, thus eliminating a variety of material from the *Bulletin*. This was approved in January 1977. He further proposed eliminating Research Announcements. Expository articles, some book reviews, and a few obituaries would make up the journal. There was no further action at the time. In April 1977, Duane Bailey, chairman of the Committee on Publication Problems, presented for the committee the following motion:

> We recommend that the new expository journal created by the Board of Trustees be called *The Bulletin of the American Mathematical Society, New Series*, and that it contain research expository articles (chosen by a collegial board of Associate Editors) and book reviews.

The motion was approved. It was noted that by implication it eliminated research announcements. When the minute was circulated, there was substantial dissatisfaction. Although the issue had been considered by the Committee on Publication Problems, the Committee to Monitor Problems in Communication, and the Executive Committee, there were Council members who thought that they had not had sufficient opportunity for discussion. The Committee on Publication Problems reconsidered its position, maintained it, and suggested the appointment of a new committee to consider the problem further. A separate journal of research announcements was considered. A moratorium on research announcements was put in place temporarily.

An ad hoc Committee on Research Announcements consisting of Felix E. Browder, chairman, Hyman Bass, Philip T. Church, R. James Milgram, Richard S. Palais, Isadore M. Singer, and Olga Taussky-Todd was appointed. It reported to the Council of 13 January 1978 with a proposal put forward by Singer, who expressed willingness to carry out the task of initiating his plan by serving as executive editor of an editorial board set up for the purpose. The proposal in brief called for a board of associate editors who, by vote of the entire board, would accept for each issue a body of research announcements from those submitted. The Council approved. This procedure turned out to be effective.

The issues of research expository papers were easily settled. The Society secured a grant from the National Science Foundation that allowed the payment of honoraria to select authors of research expository papers that somewhat facilitated the initial procurement of such articles.

The Editorial Committee of the *Bulletin, New Series,* was the committee already in place and consisted of F. E. Browder and a set of associate editors for research expository articles, P. R. Halmos for book reviews, and I. M. Singer with a set of associate editors for research announcements.

The editorial committees of the *Proceedings* and of the *Transactions* and *Memoirs* were initially small. Since 1967 members served for one term of four years but by convention were not eligible for re-election. (Prior to 1967 the term had been three years.) When additional editorial staff was needed, associate editors were appointed by the editorial committees. Associate editors had the same relation to the functioning of the journals as members of the editorial committees but were not Council members. The bylaws were amended, effective just after 1 January 1986, in such fashion that newly elected members of the editorial committees did not become members of the Council. At this point, editorial committees of the two journals were enlarged and all associate editors became members of the committees but not of the Council. In the steady state to obtain in 1990 (practically speaking in 1988), only the chairman of an editorial committee is a member of the Council. The title associate editor for these two journals then applies to an individual temporarily designated by an editorial committee to function as an editor until the Council has the opportunity to elect the person to the editorial committee in accord with the bylaws.

The Editorial Committee of the *Bulletin* continued as a committee of three, one member for each of research expository articles, research announcements, and book reviews, using boards of associate editors in the first two categories.

Members of editorial committees have always been chosen in uncontested elections. Here are the members for the *Bulletin* and *Bulletin, New Series,* for the *Proceedings,* and for the *Transactions* and *Memoirs* since 1938.

EDITORIAL COMMITTEE OF THE *Bulletin*

L. M. Graves	1/38–12/43
Paul A. Smith	1/38–12/45
Abraham Adrian Albert	1/39–12/42
Tomlinson Fort	1/39–12/44
Saunders Mac Lane	1/43–12/47
R. E. Langer	1/44–12/49
E. B. Stouffer	1/45–12/50
Deane Montgomery	1/46–12/49
G. Baley Price	1/50–12/57
W. Ted Martin	1/51–12/56

Ralph P. Boas	1/54–12/55
John C. Oxtoby	1/56–12/60
Billy Jo Pettis	1/57–12/59
Edwin E. Moise	1/58–12/63
Walter Rudin	1/60–12/65
Felix E. Browder	1/61–12/67
Edwin H. Spanier	1/64–12/66
Murray Gerstenhaber	1/66–12/71
Murray H. Protter	1/67–12/72
Gian-Carlo Rota	1/68–12/73
Hans F. Weinberger	1/71–12/77
John L. Kelley	1/73–12/75
Paul R. Halmos	1/74–12/79
Olga Taussky-Todd	1/76–12/78
Felix E. Browder	1/78–12/83
Isadore M. Singer	1/79–12/81
Meyer Jerison	1/80–12/85
Calvin C. Moore	1/82–8/85
Hyman Bass	1/84–12/86
Wu-chung Hsiang	8/85–12/87
Edgar Lee Stout	1/86–12/88
Morris W. Hirsch	1/87–
Roger E. Howe	1/88–

EDITORIAL COMMITTEE OF THE *Proceedings*

Nathan Jacobson[1]	1/50–12/53
G. A. Hedlund	1/50–12/54
A. C. Schaeffer	1/50–12/55
Richard Brauer	1/54–12/56
Shizuo Kakutani	1/55–12/57
S. S. Chern	1/56–12/56
Ralph P. Boas	1/56–12/61
Irving Kaplansky	1/57–12/59
Hans Samelson	1/57–12/60
Paul R. Halmos	1/58–12/63
Alex Rosenberg	1/60–12/65
Edwin H. Spanier	1/61–12/61
Fritz John	1/62–12/62
George Whaples	1/62–12/65
Eldon Dyer	1/62–12/67

1.Gerhard Hochschild served in place of Jacobson during 1952–53

Maurice H. Heins[2]	1/62–12/67
W. R. Wasow	1/63–12/63
Fritz John	1/64–12/66
R. Creighton Buck	1/64–12/68
W. R. Wasow[3]	1/67–12/70
William H. J. Fuchs[4]	1/68–12/71
Irving L. Glicksberg	1/68–12/70
Ernest A. Michael	1/68–12/71
P. Emery Thomas	1/68–12/71
Joseph J. Rotman	1/70–12/73
George B. Seligman	1/70–12/73
Fred G. Brauer[5]	1/71–12/74
Lee A. Rubel	1/72–12/72
Glen E. Bredon	1/71–12/75
Jacob Feldman[6]	4/71–12/75
W. Wistar Comfort	1/72–12/75
Richard R. Goldberg	1/73–12/79
Robert M. Fossum	1/74–12/77
Barbara L. Osofsky	1/74–12/77
Chandler Davis	9/74–12/76
Richard K. Miller	9/74–12/78
Thomas A. Chapman	1/76–12/79
Joseph A. Wolf	1/76–12/79
Ronald G. Douglas	1/77–12/80
David Eisenbud	1/78–12/81
Robert R. Phelps	1/78–12/81
Lawrence A. Zalcman	1/79–12/82
William E. Kirwan II	1/80–12/83
David J. Lutzer	1/80–12/83
Reinhard E. Schultz	1/80–12/83
Thomas H. Brylawski	1/80–12/84
David M. Goldschmidt	1/82–12/83
J. Jerry Uhl, Jr.	1/82–12/85
George R. Sell	1/83–12/86
Richard R. Goldberg	1/84–10/87
Irwin Kra	1/84–12/88
Daniel W. Stroock	1/84–12/87
Andrew M. Odlyzko	1/84–
Donald S. Passman	1/85–12/88

2.W.H.J. Fuchs served in place of Heins 1–9/66
3.Fred Brauer served in place of Wasow from June 1969 through February 1970
4.M. H. Heins served in place of Fuchs in January 1968
5.Richard K. Miller took over from Brauer in September 1974
6.Chandler Davis served in place of Feldman in the fall of 1974 and in 1975

Doug Curtis	1/86–7/87
Dennis Burke	5/86–
John B. Conway	5/86–
William J. Davis	5/86–
David G. Ebin	5/86–12/87
Larry J. Goldstein	5/86–12/87
Thomas J. Jech	5/86–12/88
Walter Littman	5/86–12/87
R. Daniel Mauldin	5/86–
Haynes B. Miller	5/86–12/87
Paul S. Muhly	5/86–
Bhama Srinivasan	5/86–12/87
William D. Sudderth	1/87–
Kenneth R. Meyer	1/87–
Louis J. Ratliffe, Jr.	1/87–
J. Marshall Ash	11/87–
William W. Adams	1/88–
Frederick R. Cohen	1/88–
Barbara Lee Keyfitz	1/88–
George C. Papanicolaou	1/88–
Jonathan M. Rosenberg	1/88–
James E. West	1/88–
Warren J. Wong	1/88–

EDITORIAL COMMITTEE OF THE *Transactions*[7]

William Caspar Graustein	
	1/36–12/40
Cyrus C. MacDuffee	1/37–12/42
Einar Hille	1/37–12/43
Oscar Zariski	1/41–12/47
A. A. Albert	1/43–12/48
Edward James McShane	1/44–12/46
Antoni Zygmund	1/47–12/49
G. T. Whyburn	1/48–12/52
Saunders Mac Lane[8]	1/49–12/54
L. V. Ahlfors	1/50–12/55
J. L. Doob[9]	1/50–12/55

7.*Transactions* and *Memoirs* since 1950
8.O.F.G. Schilling served in place of Mac Lane in the summer of 1954
9.Mahlon M. Day served in place of Doob from June to November 1954

Herbert Busemann[10]	1/53–12/56
Samuel Eilenberg[11]	1/55–12/59
A. H. Clifford	1/56–12/57
Mark Kac[12]	1/56–12/58
Shiing-Shen Chern	1/57–12/58
G. P. Hochschild	1/58–12/60
Lipman Bers	1/59–5/64
Isadore M. Singer	1/59–12/65
W. S. Massey	1/60–12/65
Daniel Zelinsky	1/61–12/66
M. M. Loeve	1/61–12/66
Henry P. McKean	8/62–8/63
Louis Nirenberg[13]	1/64–12/67
Dana S. Scott	1/65–12/71
Richard S. Palais	1/66–12/69
Franklin P. Peterson	1/66–12/69
David Buchsbaum	1/67–12/70
Henry P. McKean	1/67–12/70
Joseph J. Kohn	1/68–12/71
Steve Armentrout	1/70–12/73
Shlomo Sternberg	1/70–9/74
Harry Kesten	1/71–12/74
Dock S. Rim	1/71–12/74
Alistair H. Lachlan	1/72–12/75
François Trèves	1/72–12/75
Philip T. Church	1/74–12/77
Alexandra Ionescu Tulcea [Alexandra Bellow]	10/74–4/77
Stephen S. Shatz	1/75–12/78
Daniel W. Stroock	1/75–12/78
Solomon Feferman	1/76–12/79
Robert T. Seeley	1/76–12/79
W.A.J. Luxemburg	4/77–12/81
James D. Stasheff	1/78–12/81
Michael Artin	1/79–12/82
Steven Orey	1/79–12/82
R. O. Wells, Jr.	1/79–12/82
Paul H. Rabinowitz	1/80–8/82
Jan Mycielski	1/80–12/83

---

10. Leo Zippin served in place of Buseman in 1954
11. William S. Massey served in place of Eilenberg from June 1956 to September 1957
12. Joseph L. Doob served in place of Kac from January to June 1956
13. Jurgen K. Moser served in place of Nirenberg in the summer of 1965

William B. Johnson	1/82–12/85
Walter David Neumann	1/82–12/85
Joel A. Smoller	9/82–12/87
Donald L. Burkholder	1/83–12/85
Linda Preiss Rothschild	1/83–12/86
Lance W. Small	1/83–12/87
Tilla Klotz Milnor	1/84–12/87
Robert D. Edwards	1/86–4/87
Vaughan F. R. Jones	1/86–
James W. Cannon	4/87–
Ronald L. Graham	1/86–12/87
Peter W. Jones	1/86–12/88
Kenneth Kunen	1/86–12/87
Ralph L. Cohen	1/87–
Ronald Getoor	1/87–
Jerry L. Kazdan	1/87–
David J. Saltman	1/87–
Robert J. Zimmer	1/87–
James E. Baumgarten	1/88–
Roger D. Nussbaum	1/88–
Carl Pomerance	1/88–

Mathematics of Computation

Computation became of pressing importance with the advent of WWII. Computation was an art and repositories of the skills lay with diverse users, such as astronomers, actuaries, surveyors, and number theorists. There were government agencies such as Aberdeen Proving Ground (Army), Dahlgren Proving Ground (Navy), the Coast and Geodetic Survey, and the National Bureau of Standards where the art was practiced. Many mathematicians learned the skills and became interested in developing the theory.

Computation was done "by hand". That is, an algorithm or program was in the mind or the private notes of the computer, who was a person, usually female with a male supervisor. (Sometimes the understanding of the supervisor surpassed that of the computer but the skill of the latter was frequently superior.) Work was done with pencil and paper records with the aid of a motor-driven mechanical calculator, usually Monroe or Marchant. The more automated calculation of the day was done on IBM punch card machines, with packs of cards carried from one machine to another, again according to a plan in the head of the operator.

Tables were an important part of calculation. Table look-up was then easier than computing whereas direct computation by a sub-routine is now the natural route.

The journal *Mathematical Tables and Other Aids to Computation* first appeared in 1943. It was a quarterly published by the National Research Council (NRC). R. C. Archibald was the editor of volume 1 but by volume 4 he no longer served. Volumes did not correspond to calendar years until volume 5 of 1955, which was published by the National Academy of Sciences-National Research Council (NAS-NRC). Volume 56 carries the imprint of NRC alone but then NAS-NRC reappears through volume 14 of 1960. It was in that year that the name was changed to *Mathematics of Computation* corresponding to the changing nature of the field.

The issue of Society sponsorship first appeared at the Council of 28 December 1955, when E. F. Beckenbach made such a recommendation for the Committee on Printing and Publishing. A committee to investigate was authorized. However, it developed that the Division of Mathematics of the NRC wished to continue publication. When this was reported to the Council of 20 April 1956, the Council withdrew the authorization of the committee.

J. Barkley Rosser, chairman of the division, appeared before the Council of 25 January 1961 to discuss arrangements that the division was trying to make to transfer all or part of the responsibility for publication of the journal. Again a committee to study the question was authorized but there is no record of it. Instead the Committee on Printing and Publishing recommended that the Society take over the publication but not the editing of the journal. The Council demurred and authorized a committee to study the scientific merits and general objectives of the journal and to report. It consisted of Will Feller, chairman, J. L. Doob, Derrick H. Lehmer, Abraham H. Taub, and S. S. Wilks. The report of that committee stated that:

> *Mathematics of Computation* is a reputable research journal with high editorial standards. It is well on its way to becoming *the* journal in this country for publishing papers in numerical analysis and related topics.

On the basis of that report, the Planning Committee, consisting of R. Brauer, J. L. Doob, W. Feller, E. E. Floyd, J. W. Green, A. E. Meder, Jr., Lowell J. Paige, and S. Mac Lane, chairman, recommended that the Society act as publisher for the journal, with editorial responsibility remaining with the Division of Mathematics of the NRC. The recommendation carried a three year time limit. The Council of 29 August 1961 approved. The arrangement was in effect in 1961–1962. The chairman of the Editorial Committee was Harry Polachek.

At the meeting of 12 January 1964, the Council increased its operational support of the journal for the coming year and authorized a committee to study the problem of taking the journal over permanently. The Board of Trustees urged a prompt decision. The committee, consisting of A. H. Taub, chairman, E. Dyer, Harry Polachek, L. Henkin, and J. D. Swift, reported at the Council of 24 April 1964. The committee was not in agreement but several of its members were willing to propose a three year extension of the interim agreement. The Council extended the agreement for one year to give the committee additional time to deliberate.

At the Council of 25 August the committee presented three positions without full support for any of them. They were

    I Permanent continuation of the interim arrangement.
    II Transfer of the journal to the Society.
    III Return of the journal to the NRC, with qualifications.

Proposal I was characterized as "best described by the statement that a bare majority did not object to it." Opposition lay in the fact that the Society would be a mere business manager, even though the enterprise was worthy. The committee was three-to-two in favor of II, the argument against it being that it involved the Society in the publication of a specialized journal. Proposal III received the strongest support from the committee. The editors of the journal strongly favored proposal II and the Council recommended to the Trustees that it be adopted subject to the approval of the NAS-NRC.

The Board of Trustees of 17–18 December 1964 approved and so subsequently did the NAS-NRC. In 1965 the bylaws were amended in such fashion that effective 1 January 1966 *Mathematics of Computation* was an official journal of the Society and the members of the editorial committee were Council members. This journal has operated by design with a very small editorial committee, concerned with policy, and a large board of associate editors.

EDITORIAL COMMITTEE OF *Mathematics of Computation*

Eugene Isaacson	1/65–12/74
James Bramble	1/73–12/83
Alston S. Householder	1/73–12/76
John W. Wrench, Jr.	1/73–12/78
Walter Gautschi	1/76–12/86
Carl-Wilhelm de Boor	1/77–12/82
Daniel Shanks	1/78–12/84
Morris Newman	1/81–12/83
Hugh C. Williams	1/83–
Walter Gautschi	1/84–
John E. Osborn	1/84–12/86
Donald Goldfarb	1/85–

Journal of the American Mathematical Society

In 1978 the Trustees established a Committee on the Publication Program consisting of D. W. Anderson, E. Pitcher, and W. J. LeVeque, to whom were soon added M. Gerstenhaber, C. Moler, and R. J. Milgram. Following initial activity, it was relatively inactive until a rejuvenation in 1984. At a conference telephone call on 1 March 1985, the committee recommended "[t]hat a new journal be instituted, having significantly higher standards of acceptance than the *Transactions* and *Proceedings*." The membership of the committee at the time was M. Gerstenhaber, chairman, S. Armentrout, W. E. Kirwan II, W. J. LeVeque, E. Pitcher, Hugo Rossi, and Elias M. Stein. B. Janson, who was head of the publication division in the Providence office, was consultant. All but Stein were able to participate in the call. The proposal was approved by the Executive Committee and Board of Trustees with some reservations about the claim to an elite status. It was referred to Committee to Monitor Problems in Communication for study. That committee supported the idea of another inexpensive outlet for high quality papers and "agreed that the new journal could establish a very high standard only with the committed effort of one or two prestigious chief editors." The executive director was commissioned to explore the possibilities of securing editors with some persons whose names were suggested and reported favorably. The Executive Committee and Board of Trustees of 21–23 November 1985 authorized the project in principle and recommended it to the Council with the provisional title *The Journal of the American Mathematical Society*. The Council of 6 January 1986 authorized the journal with the centennial year 1988 as the date of initial publication.

The Council authorized President Irving Kaplansky to consult distinguished mathematicians about who the editors should be. The group assembled through this consultative process consisted of Michael Artin (1991) chairman, H. Blaine Lawson, Jr. (1990), Richard B. Melrose (1989), Wilfried Schmid (1989), and Robert E. Tarjan (1991). The initial terms were set to end in December of the year in parentheses but were not established until April 1987, when the group had been at work for some time and had already gathered the papers for the first issue of 1988.

The initial set of associate editors consisted of James G. Arthur, Peter J. Bickel, Gerd Faltings, Charles L. Fefferman, Michael H. Freedman, Daniel Friedan, Ronald L. Graham, Joseph Daniel Harris, H. W. Lenstra, Jr., Andrew Majda, Hugh L. Montgomery, Paul H. Rabinowitz, Karen Uhlenbeck and W. Hugh Woodin.

The recommendation from the Executive Committee and Board of Trustees to establish the journal carried with it the recommendation that there be no page charges and that recommendation was approved. This disturbed the editors of the *Proceedings* and of the *Transactions,* especially the latter. Both of these journals had page charges and had been asking to have them eliminated. The editors expressed the view that the difference in page charges could affect their journals adversely. The Trustees had previously been reluctant to eliminate page charges despite repeated requests because voluntary page charges payable from grants were the mechanism through which the National Science Foundation had agreed to support journals. The income from page charges amounted to $81,000 in 1986. The Council of 6 January 1986 had requested that page charges on all Society journals be eliminated and at its meeting of May 1987 the Trustees, with the advice of the Executive Committee, agreed.

The American Journal

There are three journals with which the Society has been involved in editorial matters, namely the *American Journal,* of which the Society was for a time joint publisher, the *Duke Mathematical Journal,* and the *Annals of Mathematics.*

The *American Journal of Mathematics* was established by Johns Hopkins University in 1878 shortly after the founding of the university itself in 1876. J. J. Sylvester was professor of mathematics at Hopkins from 1876 to 1883.

As the Society began to feel the need for a research journal in the late nineties (the *Bulletin* was then its only journal), it was concerned about the effect that a new journal might have on the already thriving *American Journal.* Thus the Society considered the possibility of acquiring the journal or participating in publication. The offer made in 1898 was one of joint publication and editing, with a financial contribution from the Society. The offer was not accepted, the difficulty, according to recollections of William Fogg Osgood, lying both with the acceptance of papers by a board of editors not under full control of the university and with the prospect of the name of the Society on the title page.

However, the *American Journal* had financial difficulties in the twenties and had shrunk greatly so that a very similar offer was accepted by the university in 1926. Joint publication began in 1927 with a board of five editors, two appointed by Johns Hopkins University and three by the Society, and with a contribution of $2500 per year by the Society.

Part of the financial arrangement with the university was that the Society had a half interest in back numbers beginning with 1927.

The *Journal* was substantially enlarged almost immediately and ran a surplus. This was an embarrasment to the Society in that the Society was subsidizing the *American Journal* and was in turn receiving a publication subsidy from the National Academy of Sciences in its role of dispensing funds of the

General Education Board. The *Journal* was then further enlarged. The study in the Council minutes of September 1945 recommended continuation of the subsidy.

In 1947 the Trustees requested a reduction of its subsidy to the *American Journal*, noting that it exceeded that of Johns Hopkins and that the journal was accumulating a surplus. President Isiah Bowman replied that the subsidy from the university was larger than it appeared in that the accounting showed neither the overhead nor the services of a part time secretary. The university both itemized the subsidy and increased it and the annual subsidy from the university and from the Society was set at $2000 for three years. It was agreed that the reserve of $5000 that had built up would be depleted. Continued Society support was conditioned on an increase in volume of publication, which had decreased from pre-WWII levels.

In 1950 the executive director reported negotiations toward a "fifty-fifty basis" with two editors each appointed by the university and the Society.

On 6 June 1950, a new agreement was made between the Society and The Johns Hopkins University. It provided among other things for ownership by the university, joint publication, at most six editors, half appointed by each participant, and business management by The Johns Hopkins University Press. The financial arrangement in the contract took account of the fact that the Society had just established institutional memberships. The dues of an institutional member were related to the amount of research published in a number of journals including the *American Journal*. The contract provided for recovery by the *American Journal* of that portion of institutional dues "to which the Journal is entitled" and, following that, a subsidy equally by the university and the Society of an amount equal to the deficit or $4000, whichever was less. There was a minor revision of the contract in 1953. The Society was also paying an annual subvention of $1500, increased in 1953 to $1650.

In 1972 the Executive Committee and Board of Trustees recommended renegotiation with a view that the Society make no financial contribution. It was the Society position that it was contributing to a journal that was running at an annual surplus while it was the position of The Johns Hopkins University that the state of affairs would change with an inevitable change of compositor and printer. The Council recommended that the agreement with The Johns Hopkins University be terminated but rescinded that action at its next meeting. However, at this point the university took over full financial responsibility.

The agreement of 1975 provides for two editors appointed by the Society subject to approval by the editor-in-chief. The *Journal* is published "with the editorial cooperation of the American Mathematical Society."

As a matter of procedure, the secretary has been securing the approval of candidates proposed by the Nominating Committee prior to their nomination by the Council as candidates in an uncontested election.

Here are the representatives of the Society on the Editorial Committee of the *American Journal* since 1938:

T. H. Hildebrandt	1/38–12/42	G. A. Hunt	1/62–12/64
Raymond Louis Wilder	1/37–12/43	George Daniel Mostow	1/63–12/68
G. D. Birkhoff	1/43–12/44	Stephen Smale	1/65–12/70
Hassler Whitney	1/44–12/49	Raoul H. Bott	1/69–12/71
L. M. Graves	1/46–12/50	Hyman Bass	1/71–5/77
Samuel Eilenberg	1/50–12/54	Isadore M. Singer	1/71–12/77
Reinhold Baer	1/51–12/56	Richard G. Swan	5/77–12/82
André Weil	1/55–4/58	Victor W. Guillemin	1/78–12/83
Harish-Chandra	1/57–4/58	Spencer J. Bloch	1/83–12/88
J. A. Dieudonné	4/58–12/61	Richard B. Melrose	1/84–12/87
Andrew M. Gleason	4/58–12/62	M. Salah Baouendi	1/88–

Duke Mathematical Journal

In 1927 and again in 1931 the Society encouraged Duke University to found a mathematics journal, noting the need for an additional periodical and the strategic position of Duke to proceed. The university did begin publication of the *Duke Mathematical Journal* in 1935. See [A], p. 17. Duke requested the Society to appoint two editors to the journal, which the Society did for forty years.

The Society did not subsidize the *Duke Journal* as it later did some other journals. In fact, the *Duke Journal* did not request a subsidy and at the same time did not plan to receive a portion of the institutional dues in the Society based on its proportional part of the page count of publication by institutional members as had the *American Journal*. Instead, the *Duke Journal* instituted a page charge of its own about 1953.

When the Society selected its editors for the *Duke Journal* in 1976 and so notified the editors and the journal, the secretary was informed that the journal wished to terminate the arrangement. The Executive Committee endorsed the response that "[t]he officers of the Society rejoice that the *Duke Journal* finds itself in a position of such strength that it no longer needs the support of the Society in supplying editors and are happy to be relieved of the obligation."

The Society representatives on the editorial board of the *Duke Mathematical Journal* since 1938 were the following:

Ø. Ore	1/38–12/40	R. H. Fox	1/56–12/58
G. T. Whyburn	1/38–12/40	Irving E. Segal	1/56–12/58
F. J. Murray	1/41–12/46	Alfred T. Brauer	1/59–12/61
Morgan Ward	1/41–12/46	Alexander D. Wallace	1/59–12/61
Gordon Pall	1/47–12/49	Richard Arens	1/61–12/64
D. J. Struik	1/47–12/49	F. B. Wright	1/61–12/64
Carl B. Allendoerfer	1/50–12/52	R. T. Prosser	1/65–12/70
R. P. Boas, Jr.	1/50–12/52	Frank L. Spitzer	1/65–12/70
R. H. Cameron	1/53–12/55	Steven Orey	1/71–12/75
S. S. Chern	1/53–12/55	Seth L. Warner	1/71–12/75

The dates given above do not represent the full term of service of these persons on the editorial board but only the years in which they represented the Society.

The Annals of Mathematics

Beginning in 1927, at the request of the editors of the *Annals*, the Society appointed associate editors, usually three at a time, for the journal. Moreover, the Society subsidized the journal in the interval 1947–1965. The subsidy began at an annual level of $1000 but increased with time to $3500. In January 1965, the editors of the *Annals* requested that the subsidy be suspended and the secretary was instructed to extend the Society's appreciation to the *Annals*. Beyond that time the Society no longer named any editors.

The Society representatives on the editorial board of the *Annals of Mathematics* since 1938 were the following:

Theophil Henry Hildebrandt		J. L. Doob	1/49–12/53
	1/40–12/45	Charles Loewner	1/55–12/57
Saunders Mac Lane	1/40–12/42	A. M. Gleason	1/55–12/60
G. T. Whyburn	1/39–12/45	J. L. Kelley	1/55–12/60
Hassler Whitney	1/43–12/45	Lars V. Ahlfors	1/58–12/60
Kurt O. Friederichs	1/46–12/48	Hans Samelson	1/61–12/66
Norman Levinson	1/46–12/53	I. E. Segal	1/61–12/66
R. L. Wilder	1/46–12/83	Daniel Zelinsky	1/61–12/66

These dates do not represent the full service of these individuals to the *Annals* but only the term during which they represented the Society. Society records show Richard Arens and Loewner as representatives in 1952–1954 but the journal does not. The inconsistencies are unresolved.

The Subsidized Journals

The initial years of a new journal are a financial drain on the publisher in that the journal must be produced and a number of copies printed equal to the estimated minimum number of subscriptions in the steady state, all before more than a handful of subscriptions have been sold. The Society has been approached repeatedly to contribute to start-up costs of new journals and also to contribute to journals in temporary financial difficulties. It has acceded to the first kind of request when the concept of the journal appeared good and the space available for publication seemed constricted. With respect to the second kind, it has been reluctant.

The subsidies to the *American Journal,* of which the Society was joint publisher for a time, and to the *Annals* have been detailed.

Other journals subsidized by the Society during various intervals between 1946 and 1976 include:

Canadian Journal of Mathematics

Historia Mathematica

Houston Journal of Mathematics

Illinois Journal of Mathematics

Journal of the Society for Industrial and Applied Mathematics

The Journal of Symbolic Logic

Michigan Mathematical Journal

Pacific Journal of Mathematics

The Rocky Mountain Journal of Mathematics

Some of these subsidies lasted only a year or two and others for nearly thirty. Since 1976 there have been none.

Meetings

The Society sponsors both general meetings and specialized conferences. It is the former that are under discussion here.

In the 1930s, the Annual Meeting, prescribed by the bylaws to take place between the 15th of December and the 15th of January of the following year, was in fact scheduled between Christmas and New Year's Day. It was held on a university campus, though accommodations may have been in hotels. Some of the meetings were in conjunction with the meetings of the American Association for the Advancement of Science. Papers in mathematics presented in Section A of AAAS were counted as presented to the Society, whether meetings were physically contiguous or not, and thus were included in the annual list, which gave references to the published paper when available. Attendance at an annual meeting was on the order of three hundred, much the same, as a percent of membership, as attendance more recently. The program consisted of the Gibbs Lecture, one or two invited addresses, perhaps a retiring presidential address or an address from Section A of AAAS, and contributed ten minute papers, numbering more than fifty. There were general sessions, with no competing events, and two or three simultaneous sections according to broad classifications of subject matter.

Summer Meetings were quite similar. They were held on university campuses with dormitory housing. They frequently followed Labor Day. The program featured the Colloquium Lectures. Some sort of expedition or picnic was a regular concomitant. There was a mandatory group photograph through 1948.

Sectional meetings were perhaps seven in number, very frequently in New York or Chicago.

Aside from the disruptions of World War II, this pattern continued, though meetings grew. The attendance at the Annual Meeting of December 1953 in Baltimore was about 600 and that of January 1960 in Washington, DC about 1300. The Annual Meeting in late January 1958 in Cincinnati signaled the change from meetings between Christmas and New Year's Day. An amendment to the bylaws had been necessary to accommodate such a change, the allowed interval now extending from 15 December to 10 February.

The style of program was altered with the Annual Meeting of 24–27 January 1963 in Berkeley. There were five Special Sessions. Each was devoted to a single topic and consisted of twenty minute papers invited by an individual selected by the Program Committee. This feature of meetings spread to the Summer and regional meetings. The sessions grew in both size and popularity until at the Annual Meeting and Summer Meeting it sometimes became necessary to limit the number of special sessions, the number of sittings of a session, and the number of papers that could be accommodated in a single session.

The Council requested a richer fare at Annual Meetings, so that in 1971 in Atlantic City there were four hour addresses. In 1972 in Las Vegas, the number was eight and that number was the norm for many years. At the same time, hour addresses no longer were free from competition with other events.

The Las Vegas location met with mixed responses from the membership. The physical arrangements of meetings, such as space and visual aids, were excellent and prices were low. However, some found the ambience of gambling intrusive and oppressive (and a handful found it too tempting). At the Business Meeting the following resolution was offered by Saunders Mac Lane:

> Whereas the circumstances in Las Vegas are not conducive to mathematical research and scholarship, be it resolved that the American Mathematical Society not meet again during this century in Las Vegas.

The secretary asked for permission not to put the resolution, if it were passed, into the public record. The motion passed by voice vote but there was a call for a count. In the meantime there was a request to make the decision public. Mac Lane withdrew the motion. It was ruled that this was not acceptable from a parliamentary standpoint. In some manner, whose parliamentary provenance is not clear, the matter was tabled. There is no record of this contretemps in the minutes of the Business Meeting.

As noted in the account of Colloquium Lectures, the number of sets of lectures in a year increased beginning in 1968.

The Annual and Summer Meetings had been five day meetings for many years, with six half days assigned to AMS and four to MAA, two days in the middle being interlaced. Beginning in 1984, four day meetings were tried, with each organization running for the entire four days and with some ground rules about choice of material in various time slots to reduce competition of subject matter. An advantage of the four day meeting was that one's expenses were reduced by one day and night of subsistence. However, it was the opinion of some that the meeting program became too crowded. Whether to return to the five day meeting was an open question as this is being written. To ameliorate the situation and in recognition of the enrichment of meeting programs from other sources, the number of hour addresses is being reduced to six.

Beginning in the Summer of 1973 in Missoula, the Society, at the behest of the Committee on Employment and Educational Policy, began to offer a Short Course prior to Summer and Annual Meetings. The first was a Preceptorial Introduction to Computer Science for Mathematicians under the direction of Jacob T. Schwartz. Short Courses were designed for persons wishing to change field or employment. Initially they were given without charge and were open to all. It soon became necessary to institute a registration fee. In some cases, particularly when computer terminals were required, it was necessary to limit enrollment.

The largest meeting of the Society was the joint Annual Meeting of January 1969 in New Orleans, where there were 4811 registrants, including 3084 members of the Society. This was the time when there was a sudden increase in the number of new Ph.D.'s and the foment in the employment market as well as the pleasant location contributed to the size of this meeting.

Meetings in the Southeast were a subject of controversy in the late 1940s. A committee consisting of Ralph Palmer Agnew, chairman, Edwin F. Beckenbach, Tomlinson Fort, G. A. Hedlund, and T. Y. Thomas was charged to study their desirability. Objection had been raised that such meetings might lead to inferior papers. The generally favorable argument was that it would encourage mathematical activity. Beckenbach argued that one who presents an unworthy paper by title is less likely to have the facts brought to light than one who presents it in person. Thomas echoed the view of Raymond L. Moore that such meetings should not be held, "that, for mathematicians on the Puerto Rican and Texan fringes of the Southeast, a Southeastern meeting would be a terrible ordeal." The Committee voted four to one in favor of such meetings. The Council of 16 April 1948 passed the following resolution:

> That the Council schedule annually southeastern meetings of the Society, the meetings to be held at whatever time may be judged most likely to encourage southeasterners to attend both southeastern and national (Summer, Christmas) meetings.

The actions of the Council to assure that the conduct of such meetings be free from racial discrimination is discussed in the section on political and social questions.

Expansion into other areas met with less controversy. At the Council of 2 September 1947 Arnold Dresden reported interest in a meeting at the University of British Columbia (UBC) in June 1948, immediately following the meeting of the Royal Society of Canada, and more generally biennial meetings of the Society in the Pacific Northwest. The invitation from UBC was accepted in principle and the general proposal was referred to Associate Secretary John Willie Green for the Far West. The Council of 30 December 1947 formally accepted the invitation from UBC and on Green's recommendation approved biennial meetings in the Pacific Northwest.

The Annual and Summer Meetings which are joint meetings with the Mathematical Association of America are not the only joint meetings of the Society. The early connection with American Association for the Advancement of Science (AAAS) has been noted but this ceased when the Annual Meeting was moved from late December to January. From time to time an Annual Meeting is joint with the Institute of Mathematical Statistics or the Association for Symbolic Logic. Other organizations meet concurrently with cooperation in arranging schedules to avoid conflicts. These include the National Council of Teachers of Mathematics, Pi Mu Epsilon, and the Association for Women in Mathematics. Sometimes regional meetings are joint meetings with a section of MAA.

Colloquium Lectures

The Colloquium Lectures date almost from the beginnings of the Society. These started as sets of lectures by Maxime Bôcher and by James Pierpont in 1896 at the third Summer Meeting in Buffalo and by William Fogg Osgood and by A. G. Webster at the fifth Summer Meeting in Cambridge. These two sets were separate from but contiguous to the Summer Meetings. The Colloquium Lectures became an established feature of Summer Meetings with the third set by O. Bolza and by E. W. Brown in Ithaca in 1901. Publication began with the fourth set of lectures by H. S. White, F. S. Woods, and E. B. van Vleck in Cambridge in 1903.

Colloquium Lectures were continued at irregular intervals of several years until 1927, when they became an annual event. The history [A] by Archibald gives a more extensive account of the origins of the Colloquium Lectures followed by a list with details about the lectures and bibliographic references in Ch VI, pp. 66–73.

There were no Colloquium Lectures in 1938, when the special program of the Semi-centennial Celebration was held, with many invited addresses.

At the Summer Meeting of 1939 in New York, there were two Colloquium Lecturers, A. A. Albert, who spoke on "Structure of Algebras," and Marshall Harvey Stone, whose title was "Convex Bodies."

The Colloquium Lectures of 1945 were held not at the Summer Meeting but at the Annual Meeting, which was held at an unusually early time in late November.

Lectures proceeded on an annual basis at Summer Meetings, excepting 1950, 1954, 1958, 1962, and 1966, when there was no Summer Meeting on account of the International Congresses.

It became apparent to the Council that the bulk of good mathematics and the number of worthy candidates to give Colloquium Lectures were increasing to the point that one set of lectures per year was insufficient to honor the speakers and cover the subject. There was a secondary phenomenon in the publication of Colloquium Lectures. Perhaps it was an acceleration of mathematical careers or of mathematical publication and perhaps the fact that candidates were being invited later in their careers, but some of the speakers had already written a definitive research monograph that, in earlier times, would have been a natural book in the Colloquium Series. In any event, the Council in 1967 decided to have two sets of Colloquium Lectures at the Summer Meeting, beginning in 1968. There was only a single set in 1970, when there was a Summer Meeting despite the almost simultaneous occurrence of the International Congress in Nice. In 1973, there were three sets of lectures, one having been added to the program of the Annual Meeting. There were two sets at the Annual Meeting of 1974, a year of no Summer Meeting because of the International Congress. The year 1975 was another year of three sets of lectures. From that point on, there has been one set of lectures at the Annual Meeting and one at the Summer Meeting, the latter usually not held in ICM years.

Here is the continuation of the list of Colloquium speakers in [A]. There were four lectures to a set except where specifically noted. When lectures were published, usually in an expanded or revised form in the series Colloquium Publications, that fact is noted. Some, such as those of Bing and Jacobson, were published long after the lectures and contain substantial subsequent developments. Otherwise the lectures are "unpublished" in the sense that substantive material appears in papers before and after the lectures but not in the Colloquium Publications. The list begins with the Twenty-first Colloquium but the numbering system fell into disuse.

Madison, University of Wisconsin, 5–8 September 1939

A. A. Albert, "Structure of Algebras."

Published as *Structure of algebras*, 1939, revised 1961.

M. H. Stone, "Convex Bodies" (3 lect.).

Hanover, Dartmouth College, 10–12 September 1940

G. T. Whyburn, "Analytic Topology."

Published as *Analytic topology*, 1941.

Chicago, University of Chicago, 2–6 September 1941
Øystein Ore, "Mathematical Relations and Structures" (3 lect.)

Poughkeepsie, Vassar College, 8–10 September 1942
R. L. Wilder, "Topology of Manifolds."
Published as *Topology of Manifolds*, 1949, revised 1963.

New Brunswick, Rutgers University, 12–13 September 1943
E. J. McShane, "Existence Theorems in the Calculus of Variations" (3 lect.).

Wellesley, Wellesley College, 13–14 August 1944
Einar Hille, "Selected Topics in the Theory of Semi-groups."
Published as *Functional analysis and semigroups* by E. Hille and Richard S. Phillips, 1948, revised 1957.

Chicago, Museum of Science and Industry, 23–24 November 1945
Tibor Radó, "Length and Area".
Published as *Length and Area*, 1948.

Ithaca, Cornell University, 20–23 August 1946
Hassler Whitney, "Topology of Smooth Manifolds."

New Haven, Yale University, 2–5 September 1947
Oscar Zariski, "Abstract Algebraic Geometry."

Madison, University of Wisconsin, 7–10 September 1948
Richard Brauer, "Representations of Groups and Rings."

Boulder, University of Colorado, 30 August–2 September 1949
G. A. Hedlund, "Topological Dynamics."
Published as *Topological dynamics* by W. H. Gottschalk and G. A. Hedlund, 1955.

Minneapolis, University of Minnesota, 4–7 September 1951
Deane Montgomery, "Topological Transformation Groups."

East Lansing, Michigan State University, 2–5 September 1952
Alfred Tarski, "Arithmetical Classes and Types of Algebraic Systems." But see *Some notions and methods on the borderline of algebra and metamathematics*, Proceedings of the International Congress of Mathematicians, 1950.

Kingston, Ontario, Queen's University and Royal Military College, 31 August–5 September 1953
Antoni Zygmund, "On the Existence and Properties of Certain Singular Integrals."

Ann Arbor, University of Michigan, 30 August–2 September 1955
Nathan Jacobson, "Jordan Algebras."

Published as *Structure and representation of Jordan algebras*, 1968.

Seattle, University of Washington, 21–24 August 1956
Salomon Bochner, "Harmonic Analysis and Probability."

University Park, Pennsylvania State University, 27–30 August 1957
N. E. Steenrod, "Cohomology Operations."

Salt Lake City, University of Utah, 1–4 September 1959
J. L. Doob, "The First Boundary Value Problem."

East Lansing, Michigan State University, 30 August–2 September 1960
S. S. Chern, "Geometrical Structures on Manifolds."

Stillwater, University of Oklahoma, 29 August–1 September 1961
George W. Mackey, "Infinite Dimensional Group Representatives."

Boulder, University of Colorado, 27–30 August 1963
Saunders Mac Lane, "Categorical Algebra."

Amherst, University of Massachusetts, 25–28 August 1964
Charles Bradford Morrey, Jr., "Multiple Integrals in the Calculus of Variations."

Ithaca, Cornell University, 31 August–3 September 1965
Alberto P. Calderón, "Singular Integrals."

Toronto, Ontario, University of Toronto, 29 August–1 September 1967
Samuel Eilenberg, "Universal Algebras and the Theory of Automata."

Madison, University of Wisconsin, 27–30 August 1968
Donald C. Spencer, "Overdetermined Systems of Partial Differential Equations."

Eugene, University of Oregon, 26–29 August 1969
Raoul H. Bott, "On the Periodicity Theorem of the Classical Groups and its Applications."
Harish-Chandra, "Harmonic Analysis on Semisimple Lie Groups."

Laramie, University of Wyoming, 25–28 August 1970
R H Bing, "Topology of 3-manifolds."
Published as *The geometric topology of 3-manifolds*, 1983

University Park, Pennsylvania State University, 31 August–3 September 1971
Lipman Bers, "Uniformization, Moduli and Kleinian Groups."
Armand Borel, "Algebraic Groups and Arithmetic Groups."

Hanover, Dartmouth College, 29 August–1 September 1972
Stephen Smale, "Applications of Global Analysis to Biology, Economics, Electrical Circuits, and Celestial Mechanics."
John T. Tate, "The Arithmetic of Elliptic Curvers."

Dallas, Fairmont Hotel, 25–28 January 1973
  Michael F. Atiyah, "The Index of Elliptic Operators."

Missoula, University of Montana, 21–24 August 1973
  Felix E. Browder, "Nonlinear Functional Analysis, and its Applications to Nonlinear Partial Differential and Integral Equations."
  Erret A. Bishop, "Schizophrenia in Contemporary Mathematics."

San Francisco, San Francisco Hilton Hotel, 15–18 January 1974
  Louis Nirenberg, "Selected Topics in Partial Differential Equations."
  John G. Thompson, "Finite Simple Groups."

Washington, Shoreham Hotel, 23–26 January 1975
  H. Jerome Keisler, "New Directions in Model Theory."

Kalamazoo, Western Michigan University, 18–22 August 1975
  Ellis R. Kolchin, "Differential Algebraic Groups."
  Elias M. Stein, "Singular Integrals, Old and New."

San Antonio, San Antonio Convention Center, 22–25 January 1976
  I. M. Singer, "Connections between Analysis, Geometry, and Topology."

Toronto, Ontario, University of Toronto, 24–27 August 1976
  Jürgen K. Moser, "Recent Progress in Dynamical Systems."

St. Louis, Chase-Park Plaza Hotel, 26–30 January 1977
  William Browder, "Differential Topology of Higher Dimensional Manifolds."

Seattle, University of Washington, 15–18 August 1977
  Herbert Federer, "Geometric Measure Theory."

Atlanta, Hyatt Regency Atlanta, 3–7 January 1978
  Hyman Bass, "Algebraic K-Theory."

Biloxi, Convention Center, 24–27 January 1979
  Philip A. Griffiths, "Complex Analysis and Algebraic Geometry."

Duluth, University of Minnesota, 22–25 August 1979
  George Daniel Mostow, "Discrete Subgroups of Lie Groups."

San Antonio, San Antonio Convention Center, 3–6 January 1980
  Wolfgang M. Schmidt, "Various Methods in Number Theory."

Ann Arbor, University of Michigan, 19–22 August 1980
  Julia Bowman Robinson, "Between Logic and Arithmetic."

San Francisco, San Francisco Hilton and Tower, 7–10 January 1981
  Mark Kac, "Some Mathematics Problems Suggested by Questions in Physics."

Pittsburgh, University of Pittsburgh, 18–21 August 1981
  Serge Lang, "Units and Class Numbers in Algebraic Geometry and Number Theory."

Cincinnati, Cincinnati Convention-Exposition Center, 13–16 January 1982
Dennis Sullivan, "Geometry, Iteration, and Group Theory."

Toronto, Ontario, University of Toronto, 23–26 August 1982
Morris W. Hirsch, "Convergence in Ordinary and Partial Differential Equations."

Denver, Denver Convention Complex, 5–8 January 1983
Charles L. Fefferman, "The Uncertainty Principle."

Albany, State University of New York, 8–11 August 1983
Bertram Kostant, "On the Coxeter Element and the Structure of the Exceptional Lie Groups."

Louisville, Commonwealth Convention Center, 25–28 January 1984
Barry Mazur, "On the Arithmetic of Curvers."

Eugene, University of Oregon, 16–19 August 1984
Paul H. Rabinowitz, "Minimax Methods in Critical Point Theory and Applications to Differential Equations."

Anaheim, Anaheim Convention Center, 9–13 January 1985
Daniel Gorenstein, "The Classification of the Finite Simple Groups."

Laramie, University of Wyoming, 12–15 August 1985
Karen K. Uhlenbeck, "Mathematical Gauge Field Theory."

New Orleans, Hyatt Regency New Orleans, 7–11 January 1986
Shing-Tung Yau, "Nonlinear Analysis."

San Antonio, San Antonio Convention Center, 21–24 January 1987
Peter David Lax, "Uses of the Non-Euclidean Wave Equation."

Salt Lake City, University of Utah, 5–8 August 1987
Edward Witten, "Mathematical Applications of Quantum Field Theory."

Atlanta, Hyatt Regency Atlanta, 6–9 January 1988
Victor W. Guillemin, "Spectral Properties of Riemannian Manifolds."

COLLOQUIUM PUBLICATIONS

Although the Colloquium Publications were set up as a vehicle for publication of Colloquium Lectures, not only have many lectures not appeared there but also the series has been open to monographs that were contributed. These include the following in addition to those noted in [A].

G. Szegö, *Orthogonal polynomals*, 1939, fourth edition 1975.

Garrett Birkhoff, *Lattice Theory*, 1940, third edition 1967.

N. Levinson, *Gap and density theorems*, 1940.

Solomon Lefschetz, *Algebraic topology*, 1942.

A. Weil, *Foundations of algebraic geometry*, 1946, revised 1962.

Joseph F. Ritt, *Differential algebra*, 1950.

Joseph Leonard Walsh, *The location of critical points of analytic and harmonic functions*, 1950.

A. C. Schaeffer and D. C. Spencer, *Coefficients of Schlicht functions*, with a chapter on "The region of values of the derivative of a Schlicht function" by A. Grad, 1950.

N. Jacobson, *Structure of rings*, 1956, revised 1964.

Ø. Ore, *Theory of graphs*, 1962.

Alfred Tarski and Steven Givant, *A formalization of set theory without variables*, 1987.

Major revisions have been noted but not reprintings, which sometimes include corrections or minor revisions.

GIBBS LECTURES

The founding of the Josiah Willard Gibbs Lectureship in 1923 is described in [A]. The lectures are a regular feature of the annual meeting, with a few omissions. They are intended to reach those members of the intellectual community with some knowledge of mathematics and interest in it and are advertised outside of the mathematical community as well as within it. When the Annual Meeting was associated with a meeting of AAAS, this effort was more effective than in recent times. The lectures are usually in some area of application of mathematics. Lecturers are frequently chosen alternately from applied mathematicians and from persons not primarily mathematicians who use mathematics.

Publication of a version of the lectures is encouraged. The lecturer has had the right to publish in the *Bulletin*. This is apparently the only such right in Society publications, though the *Bulletin* for many years appeared to accept invited addresses without refereeing. The Council of 1 September 1949 on the recommendation of R. E. Langer, speaking for the Editorial Committee of the *Bulletin*, voted that the invitation to give the Gibbs Lecture not contain assurance that the lecture would be published and that the *Bulletin* was required to publish the lecture. At the meeting of 25 April 1952 on the motion of M. H. Heins, a member of the Council by virtue of election to the Executive Committee, the Council voted that the invitation to lecture carry with it the invitation to submit a manuscript and that it not be refereed. The reasons for these changes in policy do not appear in the minutes. The lectures for 1950, 1951, and 1952 appear not to have been published.

In [A], the first fourteen lectures with place of presentation and of publication of the lectures are listed. Here is the list from that time to the present.

15. December 1939, Columbus, Ohio; Professor Theodore von Kármán, California Institute of Technology; *The engineer grapples with nonlinear problems*, Bulletin of the American Mathematical Society, v. 46 (1940), no. 8, pp. 615–683.

16. September 1941, Chicago, Illinois; Professor Sewall Wright, University of Chicago; *Statistical genetics and evolution*, Bulletin of the American Mathematical Society, v. 48 (1942), no. 4, pp. 233–246.

17. November 1943, Chicago, Illinois; Professor Harry Bateman, California Institute of Technology; *The control of elastic fluids*, Bulletin of the American Mathematical Society, v. 51 (1945), no. 9, pp. 601–646.

18. November 1944, Chicago, Illinois; Professor John von Neumann, Institute for Advanced Study; *The ergodic theorem and statistical mechanics.*

19. November 1945, Chicago, Illinois; Professor J. C. Slater, Massachusetts Institute of Technology; *Physics and the wave equation*, Bulletin of the American Mathematical Society, v. 52 (1946), no. 5, part 1, pp. 392–400.

20. November 1946, Swarthmore, Pennsylvania; Professor Subrahmanyan Chandrasekhar, University of Chicago; *The transfer of radiation in stellar atmosphere*, Bulletin of the American Mathematical Society, v. 53 (1947), no. 7, pp. 641–711.

21. December 1947, Athens, Georgia; Professor P. M. Morse, Massachusetts Institute of Technology; *Mathematical problems in operations research*, Bulletin of the American Mathematical Society, v. 54 (1948), no. 7, pp. 602–621.

22. December 1948, Columbus, Ohio; Professor Herman Weyl, Institute for Advanced Study; *Ramifications, old and new of the eigenvalue problem*, Bulletin of the American Mathematical Society, v. 56 (1950), no. 2, pp. 115–139.

23. December 1949, New York City; Professor Norbert Wiener, Massachusetts Institute of Technology; *Problems of sensory prosthesis*, Bulletin of the American Mathematical Society, v. 57 (1951), no. 1, pp. 27–35.

24. December 1950, Gainesville, Florida; Professor G. E. Uhlenbeck, University of Michigan; *Some basic problems of statistical mechanics.*

25. December 1951, Providence, Rhode Island; Professor Kurt Gödel, Institute for Advanced Study; *Some basic theorems on the foundations of mathematics and their philosophical implications.*

26. December 1952, St. Louis, Missouri; Professor Marston Morse, Institute for Advanced Study; *Topology and geometrical analysis.*

27. December 1953, Baltimore, Maryland; Professor Wassily Leontief, Harvard University; *Mathematics in economics*, Bulletin of the American Mathematical Society, v. 60 (1954), no. 3, pp. 215–233.

28. December 1954, Pittsburgh, Pennsylvania; Professor Kurt O. Friedrichs, Institute of Mathematical Sciences, New York University; *Asymptotic phenomena in mathematical physics*, Bulletin of the American Mathematical Society, v. 61 (1955), no. 6, pp. 485–504.

29. December 1955, Houston, Texas; Professor Joseph E. Meyer, University of Chicago; *The structure of simple fields*, Bulletin of the American Mathematical Society, v. 62 (1956), no. 4, pp. 332–346.

30. December 1956, Rochester, New York; Professor Marshall H. Stone, University of Chicago; *Mathematics and the future of science*, Bulletin of the American Mathematical Society, v. 63 (1957), no. 2, pp. 61–76.

31. January 1958, Cincinnati, Ohio; Professor H. J. Muller, Department of Zoology, Indiana University; *Evolution by mutation*, Bulletin of the American Mathematical Society, v. 64 (1958), no. 4, pp. 137–160.

32. January 1959, Philadelpha, Pennsylvania; Professor J. M. Burgers, University of Maryland; *On the emergence of patterns of order*, Bulletin of the American Mathematical Society, v. 69 (1963), no. 1, pp. 1–25.

33. January 1960, Chicago, Illinois; Professor Julian Schwinger, Harvard University; *Quantum field theory*.

34. January 1961, Washington, D.C.; Professor J. J. Stoker, Institute of Mathematical Sciences, New York University; *Some nonlinear problems in elasticity*, Bulletin of the American Mathematical Society, v. 68 (1962), no. 4, pp. 239–278. Published under the title *Some observations on continuum mechanics with emphasis on elasticity*.

35. January 1962, Cincinnati, Ohio; Professor C. N. Yang, Institute for Advanced Study; *Symmetry principles in modern physics*.

36. January 1963, Berkeley, California; Professor Claude E. Shannon, Massachusetts Institute of Technology; *Information theory*.

37. January 1964, Miami, Florida; Professor Lars Onsager, Yale University; *Mathematical problems of cooperative phenomena*.

38. January 1965, Denver, Colorado; Professor D. H. Lehmer, University of California, Berkeley; *Mechanical mathematics*, Bulletin of the American Mathematical Society, v. 72 (1966), no. 5, pp. 739–750.

39. January 1966, Chicago, Illinois; Professor Martin Schwarzschild, Princeton University; *Stellar evolution*.

40. January 1967, Houston, Texas; Professor Mark Kac, Rockefeller University; *Some mathematical problems in the theory of phase transitions*.

41. January 1968, San Francisco, California; Professor Eugene P. Wigner, Princeton University; *Problems of symmetry in old and new physics*, Bulletin of the American Mathematical Society, v. 74 (1968), no. 5, pp. 793–815.

42. January 1969, New Orleans, Louisiana; Professor Raymond L. Wilder, University of Michigan; *Trends and social implications of research*, Bulletin of the American Mathematical Society, v. 75 (1969), no. 5, pp. 891–906.

43. January 1970, San Antonio, Texas; Professor Walter H. Munk, Institute of Geophysics and Planetary Physics and the Scripps Institution of Oceanography, University of California, San Diego; *Tides and time.*

44. January 1971, Atlantic City, New Jersey; Professor Eberhard F. F. Hopf, Indiana University; *Ergodic theory and the geodesic flow on surfaces of constant negative curvature*, Bulletin of the American Mathematical Society, v. 77 (1971), no. 6, pp. 863–877.

45. January 1972, Las Vegas, Nevada; Professor Freeman J. Dyson, Institute for Advanced Study; *Missed opportunities*, Bulletin of the American Mathematical Society, v. 78 (1972), no. 5, pp. 635–652.

46. January 1973, Dallas, Texas; Professor Jürgen Moser, Courant Institute of Mathematical Sciences, New York University; *The stability concept in dynamical systems.*

47. January 1974, San Francisco, California; Professor Paul A. Samuelson, Massachusetts Institute of Technology, *Economics and mathematical analysis.*

48. January 1975, Washington, D.C.; Professor Fritz John, Courant Institute of Mathematical Sciences, New York University; *A priori estimates, geometric effects, and asymptotic behavior*, Bulletin of the American Mathematical Society, v. 81 (1975), no. 6, pp. 1013–1023.

49. January 1976, San Antonio, Texas; Professor Arthur S. Wightman, Princeton University; *Nonlinear functional analysis and some of its applications in quantum field theory.*

50. January 1977, St. Louis, Missouri; Professor Joseph B. Keller, Courant Institute of Mathematical Sciences, New York University; *Rays, waves, and asymptotics*, Bulletin of the American Mathematical Society, v. 84 (1978), no. 5, pp. 727–750.

51. January 1978, Atlanta, Georgia; Professor Donald E. Knuth, Stanford University; *Mathematical typography*, Bulletin of the American Mathematical Society (N.S.), v. 1 (1979), no. 2, pp. 337–372.

52. January 1979, Biloxi, Mississippi; Professor Martin Kruskal, Princeton University; *"What are solitons and inverse scattering anyway, and why should I care?"*

53. January 1980, San Antonio, Texas; Professor Kenneth Wilson, Cornell University; *The statistical continuum limit.*

54. January 1981, San Francisco, California; Professor Cathleen S. Morawetz, Courant Institute of Mathematical Sciences, New York University; *The*

mathematical approach to the sound barrier, Bulletin of the American Mathematical Society (N.S.), v. 6 (1982), no. 2, pp. 127–145. Published under the title *The mathematical approach to the sonic barrier.*

55. January 1982, Cincinnati, Ohio; Professor Elliott W. Montroll, Institute for Physical Science and Technology, University of Maryland, College Park, Maryland; Published under the title *On the dynamics and evolution of some sociotechnical systems* (edited by Bruce J. West), Bulletin of the American Mathematical Society (N.S.), v. 16 (1987), no. 1, pp. 1–46.

56. January 1983, Denver, Colorado; Professor Samuel Karlin, Stanford University, Stanford, California; *Mathematical models, problems, and controversies of evolutionary theory*, AMS Bulletin (N.S.) v. 11 (1987), No. 2, pp. 247–262.

57. January 1984, Louisville, Kentucky; Professor Herbert A. Simon, Carnegie-Mellon University, Pittsburgh, Pennsylvania; *Computer modeling of the processes of scientific and mathematical discovery.*

58. January 1985, Anaheim, California; Professor Michael O. Rabin, Harvard University, Cambridge Massachusetts and Hebrew University, Jerusalem, Israel; *Randomization in mathematics and computer science.*

59. January 1986, New Orleans, Louisiana; Professor L. E. Scriven, University of Minnesota; *The third leg: Mathematics and computation in applicable science and high technology.*

60. January 1987, San Antonio, Texas; Professor Thomas C. Spencer, Courant Institute of Mathematical Sciences, New York University; *Schrödinger operators and dynamical systems.*

61. January 1988, Pheonix; Professor David P. Ruelle, Institute des Hautes Études Scientifiques, Paris, France; *How natural is our mathematics? The example of equilibrium statistical mechanics*

The Gibbs Lecturer receives an honorarium and expenses. This was not always the case. The Trustees of 27 September 1941 agreed that in 1943 and thereafter the Gibbs Lecturer should be reimbursed for rail travel including pullman.

PRIZES

The prizes of the Society are nominally awarded by the Council. In fact, for each prize there is a small selection committee. Formerly, the committee brought a single recommendation to the Council, which approved the award. In 1962 many recurring Council duties were delegated to the Executive Committee, approval of nominations for prizes among them.

The Executive Committee became dissatisfied with this procedure. In November 1984 a subcommittee consisting of P. R. Halmos and Melvin

Hochster made a recommendation that was forwarded and approved by the Council in November. An abbreviated version of the report follows:

> Certain of the Council's decisions (the selection of the Colloquium and Gibbs lecturers and the selection of the recipients of the Society's prizes) have traditionally been delegated to special committees and then routinely approved by the EC. The procedure has not been totally satisfactory. In most cases the EC gave rubber stamp approval, sometimes on lamentably inadequate evidence.
>
> As a compromise between, in effect, doing the work of the special committees (undesirable interference) and merely being informed of their decisions (insufficient supervision), the subcommittee recommends that the EC in the future use a system of combined input and pre-approval, as follows.
>
> The special committees should submit to the EC their semifinal short list of choices (ordinarily two or three names for each slot), together with a one-page summary of the documentation justifying each name on the list, and to do so before receiving preliminary acceptance from the suggested nominees. The system should not be inflexible: if, for instance, the special committee feels strongly that there is a single obvious choice, it will suffice that that fact be communicated to the EC.
>
> The EC should then comment on each name ("acceptable", "unacceptable", "excellent"), and possibly even suggest another name or two for the special committee's consideration. The final choice is to be left to the special committee. The EC is to be informed of it, but need no longer approve it.

The new procedure has appeared to work well. The documentation coming to the Executive Committee is much more complete. It is sometimes clear initially what choice the committee will make. The Executive Committee has occasionally expressed some preferences among submitted names.

The writer is aware of no instances of the award of prizes in which recommendations were rejected and very few in which there was any dissatisfaction voiced. On the other hand, this is not the sort of question that would be documented. There have been years in which a prize was scheduled but none was awarded in that the selection committee stated that it was unable to find a suitable candidate. This was particularly true of the Steele Prizes under the earlier set of rules that governed their award.

Prizes have been a substantial feature of Annual and Summer Meetings. However the custom that a recipient lecture briefly on the work for which the prize was awarded has fallen into disuse. Responses have become expressions of thanks. Since 1979, there has been an extensive write-up of each award in the Notices, including usually a response by the recipient.

Bôcher Prize

The first prize offered by the Society was the Bôcher Prize in analysis, established through contributions amounting initially in January 1921 to $1161.79. The prize, initially of $100, was to be awarded every five years for a notable research memoir published in the *Transactions* during the previous five years by a resident of the United States or Canada. Moreover, the recipient was not to be over 40 years of age at the time of publication.

The rules have been successively modified with this and other prizes. The place of publication was changed in 1929 to any journal on which the AMS was officially represented at the time of publication and in 1935 to any recognized journal published in the United States or Canada. In 1929 the recipient was to be a member of the Society of age no more than 50 but in 1971 conditions were liberalized so that either the recipient is a member or the memoir is published in a recognized journal published in the United States or Canada. Also in 1971 the age restriction was removed. The field was restricted to analysis in 1929.

The first four awards are detailed in [A]. Here are the subsequent recipients.

Fifth award, 1938: To John von Neumann for his memoir, *Almost periodic functions and groups*. I, II, Transactions of the American Mathematical Society, volume 36 (1934), pp. 445–492, and volume 37 (1935), pp. 21–50.

Sixth award, 1943: To Jesse Douglas for his memoirs, *Green's function and the problem of Plateau*, American Journal of Mathematics, volume 61 (1939), pp. 545–589; *The most general form of the problem of Plateau*, American Journal of Mathematics, volume 61 (1939), pp. 590–608; and *Solution of the inverse problem of the calculus of variations*, Proceedings of the National Academy of Sciences, volume 25 (1939), pp. 631–637.

Seventh award, 1948: To A. C. Schaeffer and D. C. Spencer for their memoir, *Coefficients of schlicht functions*. I, II, III, IV, Duke Mathematical Journal, volume 10 (1943), pp. 611–635, volume 12 (1945), pp. 107–125, and the Proceedings of the National Academy of Sciences, volume 32 (1946), pp. 111–116, volume 35 (1949), pp. 143–150.

Eighth award, 1953: To Norman Levinson for his contributions to the theory of linear, nonlinear, ordinary, and partial differential equations contained in his papers of recent years.

Ninth award, 1959: To Louis Nirenberg for his work in partial differential equations.

Tenth award, 1964: To Paul J. Cohen for his paper, *On a conjecture of Littlewood and idempotent measures*, American Journal of Mathematics, volume 82 (1960), pp. 191–212.

Eleventh award, 1969: To I. M. Singer in recognition of his work on the index problem, especially his share in two joint papers with Michael F. Atiyah, *The index of elliptic operators*, I, III, Annals of Mathematics, Series 2, volume 87 (1968), pp. 484–530, 546–604.

Twelfth award, 1974: To Donald S. Ornstein in recognition of his paper, *Bernoulli shifts with the same entropy are isomorphic*, Advances in Mathematics, volume 4 (1970), pp. 337–352.

Thirteenth award, 1979: To Alberto P. Calderón in recognition of his fundamental work on the theory of singular integrals and partial differential equations, and in particular for his paper *Cauchy integrals on Lipschitz curves and related operators*, Proceedings of the National Academy of Sciences, USA, volume 74 (1977), pp. 1324–1327.

Fourteenth award, 1984: To Luis A. Caffarelli for his deep and fundamental work in nonlinear partial differential equations, in particular his work on free boundary problems, vortex theory and regularity theory.

Fifteenth award, 1984: To Richard B. Melrose for his solution of several outstanding problems in diffraction theory and scattering theory and for developing the analytical tools needed for their resolution.

Cole Prizes

The Cole Prizes were established initially by a gift from Frank Nelson Cole. When he retired as secretary in 1920, he was honored with a gift of $472.88 accumulated through small contributions. He in turn made this over to the Society, which decided to use it as the beginning of a prize fund. Funds accumulated and were augmented by a gift from C. A. Cole, the son of Frank Nelson Cole.

The two prizes are the Cole Prize in the Theory of Numbers and the Cole Prize in Algebra, awarded alternately every five years. The terms, except for field, are the same as those of the Bôcher Prize. The first, in algebra, was awarded to Leonard Eugene Dickson in 1928 and the second, in number theory, to H. S. Vandiver in 1931. See [A] for a more complete account.

Here are the subsequent recipients of Cole Prizes.

Third award, 1939: To A. Adrian Albert for his papers on the construction of Riemann matrices published in the Annals of Mathematics, Series 2, volume 35 (1934) and volume 36 (1935).

Fourth award, 1941: To Claude Chevalley for his paper, *La théorie du corps de classes*, Annals of Mathematics, Series 2, volume 41 (1940), pp. 394–418.

Fifth award, 1944: To Oscar Zariski for four papers on algebraic varieties published in the American Journal of Mathematics, volumes 61 (1939) and 62 (1940), and in the Annals of Mathematics, Series 2, volumes 40 (1939) and 41 (1940).

Sixth award, 1946: To H. B. Mann for his paper, *A proof of the fundamental theorem on the density of sums of sets of positive integers*, Annals of Mathematics, Series 2, volume 43 (1942), pp. 523–527.

Seventh award, 1949: To Richard Brauer for his paper, *On Artin's L-series with general group characters*, Annals of Mathematics, Series 2, volume 48 (1947), pp. 502–514.

Eighth award, 1951: To Paul Erdös for his many papers in the theory of numbers, and in particular for his paper, *On a new method in elementary number theory which leads to an elementary proof of the prime number theorem*, Proceedings of the National Academy of Sciences, volume 35 (1949), pp. 374–385.

Ninth award, 1954: To Harish-Chandra for his papers on representations of semisimple Lie algebras and groups, and particularly for his paper, *On some applications of the universal enveloping algebra of a semisimple Lie algebra*, Transactions of the American Mathematical Society, volume 70 (1951), pp. 28–96.

Tenth award, 1956: To John T. Tate for his paper, *The higher dimensional cohomology groups of class field theory*, Annals of Mathematics, Series 2, volume 56 (1952), pp. 294–297.

Eleventh award, 1960: To Serge Lang for his paper, *Unramified class field theory over function fields in several variables*, Annals of Mathematics, Series 2, volume 64 (1956), pp. 285–325; and to Maxwell A. Rosenlicht for his papers, *Generalized Jacobian varieties*, Annals of Mathematics, Series 2, volume 59 (1954), pp. 505–530, and *A universal mapping property of generalized Jacobians*, Annals of Mathematics, Series 2, volume 66 (1957), pp. 80–88.

Twelfth award, 1962: To Kenkichi Iwasawa for his paper, *Gamma extensions of number fields*, Bulletin of the American Mathematical Society, volume 65 (1959), pp. 183–226; and to Bernard M. Dwork for his paper, *On the rationality of the zeta function of an alegbraic variety*, American Journal of Mathematics, volume 82 (1960), pp. 631–648.

Thirteenth award, 1965: To Walter Feit and John G. Thompson for their joint paper, *Solvability of groups of odd order*, Pacific Journal of Mathematics, volume 13 (1963), pp. 775–1029.

Fourteenth award, 1967: To James B. Ax and Simon B. Kochen for a series of three joint papers, *Diophantine problems over local fields*, I, II, III,

American Journal of Mathematics, volume 87 (1965), pp. 605–630, 631–648, and Annals of Mathematics, Series 2, volume 83 (1966), pp. 437–456.

Fifteenth award, 1970: To John R. Stallings for his paper, *On torsion-free groups with infinitely many ends*, Annals of Mathematics, Series 2, volume 88 (1968), pp. 312–334; and to Richard G. Swan for his paper, *Groups of cohomological dimension one*, Journal of Algebra, volume 12 (1969), pp. 585–610.

Sixteenth award, 1972: To Wolfgang M. Schmidt for the following papers: *On simultaneous approximation of two algebraic numbers by rationals*, Acta Mathematica (Uppsala), volume 119 (1967), pp. 27–50; *T-numbers do exist*, Symposia Mathematica, volume IV, Academic Press, 1970, pp. 1–26; *Simultaneous approximation to algebraic numbers by rationals*, Acta Mathematica (Uppsala), volume 125 (1970), pp. 189–201; *On Mahler's T-numbers*, Proceedings of Symposia in Pure Mathematics, volume 20, American Mathematical Society, 1971, pp. 275–286.

Seventeenth award, 1975: To Hyman Bass for his paper, *Unitary algebraic K-theory*, Springer Lecture Notes in Mathematics, volume 343, 1973; and to Daniel G. Quillen for his paper, *Higher algebraic K-theories*, Springer Lecture Notes in Mathematics, volume 341, 1973.

Eighteenth award, 1977: To Goro Shimura for his two papers, *Class fields over real quadratic fields and Heche operators*, Annals of Mathematics, Series 2, volume 95 (1972), pp. 130–190; and *On modular forms of half integral weight*, Annals of Mathematics, Series 2, volume 97 (1973), pp. 440–481.

Nineteenth award, 1980: To Michael Aschbacher for his paper, *A characterization of Chevalley groups over fields of odd order*, Annals of Mathematics, Series 2, volume 106 (1977), pp. 353–398; and to Melvin Hochster for his paper *Topics in the homological theory of commutative rings*, CBMS Regional Conference Series in Mathematics, Number 24, American Mathematical Society, 1975.

Twentieth award, 1982: To Robert P. Langlands for pioneering work on automorphic forms, Eisenstein series and product formulas, particularly for his paper *Base change for* GL(2), Annals of Mathematics Studies, volume 96, Princeton University Press, 1980; and to Barry Mazur for outstanding work on elliptic curves and Abelian varieties, especially on rational points of finite order, and his paper *Modular curves and the Eisenstein ideal*, Publications Mathématiques de l'Institut des Hautes Études Scientifiques, volume 47 (1977), pp. 33–186.

Twenty-First award, 1985: To George Lusztig for his fundamental work on the representation theory of finite groups of Lie type. In particular for his contributions to the classification of the irreducible representations in characteristic zero of the groups of rational points of reductive groups over finite fields, appearing in *Characters of Reductive Groups Over Finite Fields*, Annals of Mathematics Studies, volume 107, Princeton University Press, 1984.

Twenty-Second award, 1987: To Dorian M. Goldfeld for his paper, *Gauss's class number problem for imaginary quadratic fields*, Bulletin of the American Mathematical Society, volume 13, (1985), pp. 23–37; and to Benedict H. Gross and Don B. Zagier for their paper, *Heegner points and derivatives of L-series*, Inventiones Mathematicae, volume 84 (1986), pp. 225–320.

Veblen Prize

The Oswald Veblen Prize in Geometry was instituted through contributions from former students and colleagues augmented by a gift from his widow that brought the fund to $2000. It is specified that the award shall be made in geometry or topology under the same conditions as those set for the Bôcher Prize. After the prize was in place it was awarded every five years.

Here is the list of winners.

First award, 1964: To C. D. Papakyriakopoulos for his papers, *On solid tori*, Annals of Mathematics, Series 2, volume 66 (1957), pp. 1-26, and *On Dehn's lemma and the asphericity of knots*, Proceedings of the National Academy of Sciences, volume 43 (1957), pp. 169–172

Second award, 1964: To Raoul Bott for his papers, *The space of loops on a Lie group*, Michigan Mathematical Journal, volume 5 (1958), pp. 35–61, and *The stable homotopy of the classical groups*, Annals of Mathematics, Series 2, volume 70 (1959), pp. 313–337.

Third award, 1966: To Stephen Smale for his contributions to various aspects of differential topology.

Fourth award, 1966: To Morton Brown and Barry Mazur for their work on the generalized Schoenflies theorem.

Fifth award, 1971: To Robion C. Kirby for his paper, *Stable homeomorphisms and the annulus conjecture*, Annals of Mathematics, Series 2, volume 89 (1969), pp. 575–582.

Sixth award, 1971: To Dennis P. Sullivan for his work on the Hauptvermutung summarized in the paper, *On the Hauptvermutung for manifolds*, Bulletin of the American Mathematical Society, volume 73 (1967), pp. 598–600.

Seventh award, 1976: To William P. Thurston for his work on foliations.

Eighth award, 1976: To James Simons for his work on minimal varieties and characteristic forms.

Ninth award, 1981: To Mikhael Gromov for his work relating topological and geometric properties of Riemannian manifolds.

Tenth award, 1981: To Shing-Tung Yau for his work in nonlinear partial differential equations, his contributions to the topology of differentiable manifolds, and for his work on the complex Monge-Ampère equation on compact complex manifolds.

Eleventh award, 1986: To Michael H. Freedman for his work in differential geometry and, in particular, the solution of the four-dimensional Poincaré conjecture.

BIRKHOFF PRIZE AND WIENER PRIZE

The George David Birkhoff Prize in Applied Mathematics and the Norbert Wiener Prize in Applied Mathematics were established at the same time. The initial contribution for the former came from the Birkhoff family and for the latter from the Department of Mathematics of the Massachusetts Institute of Technology. Each is to be awarded for an outstanding contribution to "applied mathematics in the highest and broadest sense." Whereas the Bôcher, Cole, and Veblen prizes are for a specific memoir, it is recognized that work in applied mathematics is sometimes developmental and accumulative, so that the award is for a line of distinguished work rather than a pinpointed paper.

These two prizes are awarded jointly by AMS and SIAM. The recipient must be a member of one of the two societies and a resident of the United States, Canada, or Mexico. Ordinarily the Birkhoff Prize is awarded at an AMS meeting and the Wiener Prize at a SIAM meeting.

Here are the recipients of the Birkhoff Prize.

First award, 1968: To Jürgen K. Moser for his contributions to the theory of Hamiltonian dynamical systems, especially his proof of the stability of periodic solutions of Hamiltonian systems having two degrees of freedom and his specific applications of the ideas in connection with this work.

Second award, 1973: To Fritz John for his outstanding work in partial differential equations, in numerical analysis, and, particularly, in nonlinear elasticity theory; the latter work has led to his study of quasi-isometric mappings as well as functions of bounded mean oscillation, which have had impact in other areas of analysis.

Third award, 1973: To James N. Serrin for his fundamental contributions to the theory of nonlinear partial differential equations, especially his work on existence and regularity theory for nonlinear elliptic equations, and applications of his work to the theory of minimal surfaces in higher dimensions.

Fourth award, 1978: To Garrett Birkhoff for bringing the methods of algebra and the highest standards of mathematics to scientific applications.

Fifth award, 1978: To Mark Kac for his important contributions to statistical mechanics and to probability theory and its applications.

Sixth award, 1978: To Clifford A. Truesdell for his outstanding contributions to our understanding of the subjects of rational mechanics and nonlinear materials, for his efforts to give precise mathematical formulation to these classical subjects, for his many contributions to applied mathematics in the fields of acoustic theory, kinetic theory, and nonlinear elastic theory, and the thermodynamics of mixtures, and for his major work in the history of mechanics.

Seventh award, 1983: To Paul R. Garabedian for his important contributions to partial differential equations, to the mathematical analysis of problems of transonic flow and airfoil design by the method of complexification, and to the development and application of scientific computing to problems of fluid dynamics and plasma physics.

The winners of the Wiener Prize have been the following:

First award, 1970: To Richard E. Bellman for his pioneering work in the area of dynamic programming, and for his related work on control, stability, and differential-delay equations.

Second award, 1975: To Peter D. Lax for his broad contributions to applied mathematics, in particular, for his work on numerical and theoretical aspects of partial differential equations and on scattering theory.

Third award, 1980: To Tosio Kato for his distinguished work in the perturbation theory of quantum mechanics.

Fourth award, 1980: To Gerald B. Whitham for his broad contributions to the understanding of fluid dynamical phenomena and his innovative contributions to the methodology through which that understanding can be constructed.

Fifth award, 1985: To Clifford S. Gardner for his contributions to applied mathematics in the areas of supersonic aerodynamics, plasma physics and hydromagnetics, and especially for his contributions to the truly remarkable development of inverse scattering theory for the solution of nonlinear partial differential equations.

STEELE PRIZES

The will of Leroy P. Steele bequeathed the residue of his estate to the Society. The capital value turned out to be about $145,000 and there were some items of income for a few years in addition. The income from the fund was to be used for "a prize or prizes for outstanding published mathematical research at such intervals and for such types of mathematical research as the Council of the Society shall in its sole discretion determine." It was further suggested but not required that prizes be named in memory of George D. Birkhoff, William Fogg Osgood, and William Caspar Graustein. The Council did in fact name the prizes the Leroy P. Steele Prizes in honor of the three named mathematicians. The names of Birkhoff, Osgood, and Graustein are regularly mentioned in the announcements.

The existence of the bequest and its approximate value were known some time before the money became available. The Council in 1967–1968 was considering the establishment of a research expository journal, which in fact did not materialize. As an interim measure, which in fact became "permanent," the publication of research expository papers in the *Bulletin* was encouraged. With the prospect of the Steele Fund, the Council authorized the president to appoint a Steele Prize Committee to examine unsolicited articles for the *Bulletin* with the view of awarding prizes and publishing the articles. There was some question whether this plan would meet the requirements of the Steele bequest.

The committee consisted of F. Browder, S. S. Chern, M. Gerstenhaber, Edwin Hewitt, N. Jacobson, S. Lang, J. Milnor, and A. Seidenberg. Gerstenhaber, who was editor of the *Bulletin* for expository addresses, was chairman. The plan as they developed it differed somewhat from the initial proposal. It was adopted by the Council, which agreed to award prizes for outstanding mathematical research, with most favorable consideration given to papers distinguished for their exposition and covering broad areas of mathematics. Beyond the first award to Solomon Lefschetz in 1970, there were several prizes a year.

The result was not completely satisfactory as evidenced by the fact that there were no awards in 1973, 1974, 1976, 1977, and 1978. At the same time that it recommended in August 1975 that no prize be awarded in 1976, the Steele Prize Committee proposed a change in the prize. The committee expressed the view that the then current plan for the prizes did not emphasize research to the degree stipulated in the Steele will and was not effective in promoting high quality exposition. The committee recommended replacing the current procedure by a plan formulated by P. R. Halmos calling for "a large prize for a spectacular piece of research to be given on the rare occasions... when such a piece of research comes to light." Although the committee supported the proposal by a vote of 6-1, it failed in the Council.

In 1976, Hans Weinberger, chairman of the Steele Prize Committee, moved to study the Halmos plan and related questions but the Council voted simply to have an ad hoc committee to consider the future of the Steele Prize. The Committee on Prizes, consisting of James B. Serrin, Chairman, Walter Feit, Phillip A. Griffiths, P. Halmos, Victor L. Klee, P. Lax, and J. Milnor, made recommendations that were approved in the following form by the Council of 16 April 1977. There should be three prizes awarded annually in the following categories:

(1) for the cumulative influence of the total mathematical work of the recipient, high level of research over a period of time, particular influence

on the development of a field, and influence on mathematics through Ph.D. students;

(2) for a book or substantial survey or expository-research paper;

(3) for a paper, whether recent or not, which has proved to be of fundamental or lasting importance in its field, or a model of important research.

The first prizes under the new formulation were awarded in January 1979.

Here are the Steele Prize awards.

August 1970: To Solomon Lefschetz for his paper, *A page of mathematical autobiography*, Bulletin of the American Mathematical Society, volume 74 (1968), pp. 854–879.

August 1971: To James B. Carrell for his paper, written jointly with Jean A. Dieudonné, *Invariant theory, old and new*, Advances in Mathematics, volume 4 (1970), pp. 1–80.

August 1971: To Jean A. Dieudonné for his paper, *Algebraic geometry*, Advances in Mathematics, volume 3 (1969), pp. 223–321, and for his paper, written jointly with James B. Carrell, *Invariant theory, old and new*, Advances in Mathematics, volume 4 (1970), pp. 1–80.

August 1971: To Phillip A. Griffiths for his paper, *Periods of integrals on algebraic manifolds*, Bulletin of the American Mathematical Society, volume 76 (1970), pp. 228–296.

August 1972: To Edward B. Curtis for his paper, *Simplicial homotopy theory*, Advances in Mathematics, volume 6 (1971), pp. 107–209.

August 1972: To William J. Ellison for his paper, *Waring's problem*, American Mathematical Monthly, volume 78 (1971), pp. 10–36.

August 1972: To Lawrence F. Payne for his paper, *Isoperimetric inequalities and their applications*, SIAM Review, volume 9 (1967), pp. 453–488.

August 1972: To Dana S. Scott for his paper, *A proof of the independence of the continuum hypothesis*, Mathematical Systems Theory, volume 1 (1967), pp. 89–111.

January 1975: To Lipman Bers for his paper, *Uniformization moduli, and Kleinian groups*, Bulletin of the London Mathematical Society, volume 4 (1972), pp. 257–300.

January 1975: To Martin D. Davis for his paper, *Hilbert's tenth problem is unsolvable*, American Mathematical Monthly, volume 80 (1973), pp. 233–269.

January 1975: To Joseph L. Taylor for his paper, *Measure algebras*, CBMS Regional Conference Series in Mathematics, Number 16, American Mathematical Society, 1972.

August 1975: To George W. Mackey for his paper, *Ergodic theory and its significance for statistical mechanics and probability theory*, Advances in Mathematics, volume 12 (1974), pp. 178–286.

August 1975: To H. Blaine Lawson for his paper, *Foliations*, Bulletin of the American Mathematical Society, volume 80 (1974), pp. 369–418.

1976, 1977, 1978: No awards were made.

January 1979: To Salomon Bochner for his cumulative influence on the fields of probability theory, Fourier analysis, several complex variables, and differential geometry.

January 1979: To Hans Levy for three fundamental papers: *On the local character of the solutions of an atypical linear differential equation in three variables and a related theorem for regular functions of two complex variables*, Annals of Mathematics, Series 2, volume 64 (1956), pp. 514–522; *An example of a smooth linear partial differential equation without solution*, Annals of Mathematics, Series 2, volume 66 (1957), pp. 155–158; *On hulls of holomorphy*, Communications in Pure and Applied Mathematics, volume 13 (1960), pp. 587–591.

August 1979: To Antoni Zygmund for his cumulative influence on the theory of Fourier series, real variables, and related areas of analysis.

August 1979: To Robin Hartshorne for his expository research article *Equivalence relations on algebraic cycles and subvarieties of small codimension*, Proceedings of Symposia in Pure Mathematics, volume 29, American Mathematical Society, 1975, pp. 129–164; and his book *Algebraic geometry*, Springer-Verlag, Berlin and New York, 1977.

August 1979: To Joseph J. Kohn for his fundamental paper: *Harmonic integrals on strongly convex domains*, I, II, Annals of Mathematics, Series 2, volume 78 (1963), pp. 112–248 and volume 79 (1964), pp. 450–472.

August 1980: To André Weil for the total effect of his work on the general course of twentieth century mathematics, especially in the many areas in which he has made fundamental contributions.

August 1980: To Harold M. Edwards for mathematical exposition in his books *Riemann's zeta function*, Pure and Applied Mathematics, number 58, Academic Press, New York and London, 1974; and *Fermat's last theorem*, Graduate Texts in Mathematics, number 50, Springer-Verlag, New York and Berlin, 1977.

August 1980: To Gerhard P. Hochschild for his significant work in homological algebra and its applications.

August 1981: To Oscar Zariski for his work in algebraic geometry, especially his fundamental contributions to the algebraic foundations of this subject.

August 1981: To Eberhard Hopf for three papers of fundamental and lasting importance: *Abzweigung einer periodischen Lösung von einer stationären Lösung eines Differential systems*, Berichte über die Verhandlungen der Sächsischen Akademie der Wissenschaften zu Leipzig. Mathematisch-Naturwissenschaftliche Klasse, volume 95 (1943), pp. 3–22; *A mathematical example displaying features of turbulence*, Communications on Applied Mathematics, volume 1 (1948), pp. 303–322; and *The partial differential equation* $u_t + uu_x = \mu u_{xx}$, Communications on Pure and Applied Mathematics, volume 3 (1950), pp. 201–230.

August 1981: To Nelson Dunford and Jacob T. Schwartz for their expository book, *Linear operators*, Part I, *General theory*, 1958; Part II, *Spectral theory*, 1963; Part III, *Spectral operators*, 1971, Interscience Publishers, New York.

August 1982: To Lars V. Ahlfors for his expository work in *Complex analysis* (McGraw-Hill Book Company, New York, 1953), and in *Lectures on quasiconformal mappings* (D. Van Nostrand Co., Inc., New York, 1966) and *Conformal invariants* (McGraw-Hill Book Company, New York, 1973).

August 1982: To Tsit-Yuen Lam for his expository work in his book *Algebraic theory of quadratic forms* (1973), and four of his papers: K_0 *and* K_1—*an introduction to algebraic K-theory* (1975), *Ten lectures on quadratic forms over fields* (1977), *Serre's conjecture* (1978), and *The theory of ordered fields* (1980).

August 1982: To John W. Milnor for a paper of fundamental and lasting importance, *On manifolds homeomorphic to the 7-sphere*, Annals of Mathematics (2) **64** (1956), pp. 399–405.

August 1982: To Fritz John for the cumulative influence of his total mathematical work, high level of research over a period of time, particular influence on the development of a field, and influence on mathematics through Ph.D. students.

August 1983: To Paul R. Halmos for his many graduate texts in mathematics and for his articles on how to write, talk and publish mathematics.

August 1983: To Steven C. Kleene for three important papers which formed the basis for later developments in generalized recursion theory and descriptive set theory: *Arithmetical predicates and function quantifiers*, Transactions of the American Mathematical Society 79 (1955), pp. 312–340; *On the forms of the predicates in the theory of constructive ordinals (second paper)*, American Journal of Mathematics 77 (1955), pp. 405–428; and *Hierarchies of number-theoretic predicates*, Bulletin of the American Mathematical Society 61 (1955), pp. 193–213.

August 1983: To Shiing-Shen Chern for the cumulative influence of his total mathematical work, high level of research over a period of time, particular influence on the development of the field of differential geometry, and influence on mathematics through Ph.D. students.

August 1984: To Elias M. Stein for his book, *Singular integrals and the differentiability properties of functions*, Princeton University Press (1970).

August 1984: To Lennart Carleson for his papers: *An interpolation problem for bounded analytic functions*, American Journal of Mathematics, volume 80 (1958), pp. 921–930; *Interpolation by bounded analytic functions and the Corona problem*, Annals of Mathematics (2), volume 76 (1962), pp. 547–559; and *On convergence and growth of partial sums of Fourier series*, Acta Mathematica volume 116 (1966), pp. 135–157.

August 1984: To Joseph L. Doob for his fundamental work in establishing probability as a branch of mathematics and for his continuing profound influence on its development.

August 1985: To Michael Spivak for his five-volume set, *A Comprehensive Introduction to Differential Geometry* (second edition, Publish or Perish, 1979).

August 1985: To Robert Steinberg for three papers on various aspects of the theory of algebraic groups: *Representations of algebraic groups*, Nagoya Mathematical Journal, volume 22 (1963), pp. 33–56; *Regular elements of semisimple algebraic groups*, Institut des Hautes Études Scientifiques, Publications
Mathématiques, volume 25 (1965), pp. 49–80; and *Endomorphisms of linear algebraic groups*, Memoirs of the American Mathematical Society, volume 80 (1968).

August 1985: To Hassler Whitney for his fundamental work on geometric problems, particularly in the general theory of manifolds, in the study of differentiable functions on closed sets, in geometric integration theory, and in the geometry of the tangents to a singular analytic space.

January 1986: To Donald E. Knuth for his expository work, *The Art of Comnputer Programming*, 3 Volumes (1st Edition 1968, 2nd Edition, 1973).

January 1986: To Rudolf E. Kalman for his two fundamental papers: *A new approach to linear filtering and prediction problems*, Journal of Basic Engineering, volume 82, (1960), pp. 35–45; and *Mathematical description of linear dynamical systems*, SIAM Journal on Control and Optimization, volume 1 (1963), pp. 152–192; and for his contribution to a third paper, (with R. S. Bucy) *New results in linear filtering and prediction theory*, Journal of Basic Engineering, volume 83D (1961), pp. 95–108.

January 1986: To Saunders MacLane for his many contributions to algebra and algebraic topology, and in particular for his pioneering work in homological and categorical algebra.

August 1987: To Martin Gardner for his many books and articles on mathematics and particularly for his column "Mathematical games" in *Scientific American*.

August 1987: To Herbet Federer and Wendell Fleming for their pioneering paper, *Normal and integral currents*, Annals of Mathematics, volume 72 (1960), pp. 458–520.

August 1987: To Samuel Eilenberg for his fundamental contributions to topology and algebra, in particular for his classic papers on singular homology and his work on axiomatic homology theory which had a profound influence on the development of algebraic topology.

As has been noted, the Bôcher Prize was initially in the amount of $100. All of the prizes of the Society were kept at the same level, which was increased to $200 in 1954. With the advent of the Steele bequest the Steele Prize was initially $1000. When the terms of the Steele Prize were revised, two changes were made concomitantly. Steele Prizes were set at $1500 and it was agreed to augment each of the other prizes of the Society to a total of $1500 using income from the Steele Prize Fund.

The Steele Prize Fund continued to accumulate faster than it was being disbursed and could be used only for prizes. The Council has always been reluctant to have too many prizes and to subdivide mathematics too finely. Thus the additional funds were used in two ways. Over an interval of about five years, roughly 1980 to 1984, two prizes rather than one were awarded in a field. Second, beginning in 1984 the amount of each prize was set at $4000.

Sectional Meetings

In addition to the Annual and Summer Meetings, the Society has conducted several one or two day meetings each year. The pattern of these has changed very little, consisting of invited hour addresses and contributed papers. The number of hour addresses has increased with time from one or two to three or four. When the concept of Special Sessions appeared, it was introduced at sectional meetings as well. Arrangements are made by the cognizant associate secretary, usually in cooperation with a local department. There is staff support through the Providence office but it does not extend to supplying personnel at the meeting.

Contributed Papers

The purpose and use of contributed papers have changed. In the original model one did a piece of research, wrote a paper, and submitted it to a journal. At the same time one contributed the paper by speaking about it or offering it by title. The paper was reported by an abstract, ostensibly prepared by the secretary or associate secretary in the beginning but later written by the contributor. There was provision for the possibility that the work was not ready for submission to a journal in that the abstract might be labeled "preliminary report." When the paper appeared, the bibliographic information was printed in the *Bulletin* up to the point that *Mathematical Reviews* came into existence.

In fact, preliminary reports became a more prominent feature and were frequently not so tagged. Moreover, many abstracts appeared that corresponded to no future paper. There were two reasons. It came to pass that many institutions would pay at least part of the travel expenses to a meeting for one presenting a paper. The abstract became a ticket. The pressure to publish caused people to contribute papers. The abstract became a form of publication, albeit not refereed.

Although abstracts are not refereed, they are screened. The bylaw states:

> Papers intended for presentation at any meeting of the Society shall be passed upon in advance by a program committee appointed by or under the authority of the Council; and only such papers shall be presented as shall have been approved by such committee.

The committee is the Program Committee in the case of national meetings and the appropriate Committee to Select Hour Speakers in the case of regional meetings. In fact, the cognizant associate secretary does the work, occasionally with the help of a referee or even less frequently by formal referral to the committee. There is occasional serious mathematics that is recognizably in error but an abstract of sixteen lines of eighty characters is usually too brief for an error to be evident. The problems are rare and usually consist of abstracts that are mathematically empty, either mathematical vocabulary without meaning or the promotion of a prejudice or a cause.

Costs of Meetings

The pattern of paying for the costs of meetings has changed. It was once the case that a major part of the cost was the unrecorded and unrecognized cost that fell on the host institution.

At the meeting of 24 April 1953, Secretary Edward Griffith Begle reported that "there are some institutions which are not able to make provisions for certain [expenses incurred by host institutions], such as janitorial expenses, and consequently they are borne by the individual members of the host department. This has resulted in the Society's receiving fewer and less frequent invitations." He requested and the Council approved authorization to approve payments of bills for meeting expenses for janitorial fees, printing, etc. but not for entertainment, up to a maximum of $25 for a one day meeting, $50 for a two day meeting, and $100 for a Summer or Annual Meeting, amounting to a maximum of $600 per year. Experiments in producing programs by a less expensive method than letterpress suggested that the cost of programs could be decreased by at least $600 per year, so that the total cost of meetings was expected not to increase.

More recently there has been no ceiling except prudence on expenditure for meetings. Registration fees were set with the intent of recovering costs. In fact the Annual Meeting sometimes has a surplus and the Summer Meeting usually has a deficit, the latter frequently larger than the former.

Cost of sectional meetings is not a problem unless one encounters rules requiring a substantial custodial and operating staff at overtime rates.

Book exhibits (and recently computer hardware and software exhibits) have been an increasing source of income at Summer and (especially) Annual Meetings, serving to offset some of the expenses. There has always been a rule that exhibits should be of interest to registrants as mathematicians. At the meeting of 24 April 1953, the Council acceded to a request from the MAA that a restriction on book exhibitors that they show only books at the graduate level or higher be lifted.

REGISTRATION FEES

In the early days of the Society, meetings were handled almost entirely by volunteer help, consisting of the officers, particularly the cognizant associate secretary, and representatives of the host institution. The host institution bore some costs, for example the loan of a department secretary, and supplied classrooms. When hotels were suggested there was no reservation service offered. In this ambience there was little need for registration fees. Meetings held in conjunction with AAAS had registration fees of $1.00 in the 1920s and early 1930s. The first observed registration fee at a Society meeting was $0.50 in September 1934 at Williams College. There was a registration fee of $2.00 at the Semicentennial.

Several factors contributed to the institution and increase in registration fees. More services, such as the handling of room reservations in dormitories and hotels, were offered. Better service, such as preregistration, which obviated waiting in line, was provided. Universities turned to their facilities as a source of income and charged both rent for meeting rooms and conference fees. Not only hotels but also convention centers were used as meeting sites and the latter had fees for service.

Summer Meetings

In 1984 the issue was raised by Susan J. Friedlander about whether to continue to hold Summer Meetings. Many arguments on both sides were presented. Summer Meetings lose money, which was a reason the problem was discussed. On the other hand the Summer and Annual Meetings together might break even. The fact that Summer Meetings are of interest to and positively desired by the Mathematical Association of America, with whom they are jointly held, is significant. They serve a different segment of membership from other meetings. They serve substantial scientific purposes, for example the Colloquium Lectures, which might otherwise be lost. On the other hand, the pattern of meetings in general has changed greatly with time, the increase in conferences being an important factor.

In 1986 it was decided to continue to hold Summer Meetings in the indefinite future. The cost sharing with the association had been done by a formula that at one time was fair but that through changing circumstances put the Society at a disadvantage. It was readjusted to equitable terms.

Employment Register

The Council of 9 December 1953 considered a report from the Committee on the Employment Register, consisting of Leon W. Cohen, chairman, and J. A. Clarkson. The committee had consulted with the corresponding committee of the Mathematical Association of America, of which J. S. Frame was chairman. The recommendation was "that the Society undertake to establish a placement service for its members and, in cooperation with interested mathematical and statistical groups, for nonmembers concerned with mathematics and its applications." Various potential difficulties and experiences of related organizations were noted. "It is very likely that any placement agency sponsored by the Society would need to charge fees for its service." "One important consideration, in any feasibility study of the proposed placement service, is its value as an inducement for prospective sustaining [i.e., institutional] members. The schedule of placement fees could be made slightly lower for members... and thus serve to attract both individuals and organizations to seek membership in the Society."

It was the opinion of the Council than an employment register receiving applications from individuals would become a very large operation and thus the Council approved a register consisting only of a file of academic institutions, industrial firms, and government agencies, available at Summer, Annual, and April meetings. Contacts between applicants and employers were then to be the responsibility of the former. The Trustees were initially insistent on going no further in order to control costs. The stricture did not hold.

A joint AMS-MAA Committee on Employment Opportunities was appointed to operate the register. It consisted of J. S. Frame, chairman, H. M. Bacon, T. R. Hollcroft, M. Ostrofsky, and J. A. Ward. This committee replaced ad hoc arrangements for supervision of the register.

In April 1955 the Council authorized the increase in the committee to seven, with two members representing the Society for Industrial and Applied Mathematics, namely G. W. Patterson and R. Berkowitz.

In August 1956 the Council authorized the committee to prepare a list of mathematicians about to retire. The intent was to alleviate the teacher shortage by effective placement of these persons.

At the Council of 28 January 1958, the committee was discharged with special thanks to Frame. This coincided with his resignation in order to take a leave. The operation of the register became a staff function of the Providence office for the three cooperating organizations. However, an Employment Register Supervisory Committee consisting initially of W. M. Hirsch, chairman and AMS representative, A. E. Taylor, MAA representative, and R. F. Rinehart, SIAM representative, was appointed to handle policy questions. Expenses of operating the register were to be divided.

At the Council of 23 January 1967, it was noted that positions available at institutions censured by the American Association of University Professors had occasionally been listed. After study of the problem the Council of 7 April 1967 agreed to continue with such listings but to footnote them, with the advance knowledge of the employer, as coming from censured administrations. Later the list of censured administrations was published regularly.

The journal *Employment Information for Mathematicians*, later called *Employment Information in the Mathematical Sciences*, was a separate venture that began publication in 1970. It regularly listed open positions. When the legal obligation to advertise positions to comply with legislation about nondiscrimination became apparent, the following policy statement was adopted:

> The Council of the AMS adopts the principles that all positions in the mathematical sciences shall insofar as practicable be advertised, and that the standard place for advertisements to appear is in the publication *Employment Information for Mathematicians*.

The matching of applicants and employers to suit in substantial part the wishes of each in an effective scheduling of job interviews at the January meeting became a formidable problem as numbers increased and the system gained credibility. The Mathematical Sciences Employment Register apparently has a unique service. It is dependent on a matching program based on an algorithm of Donald R. Morrison. See his paper "Matching Algorithms" in *Journal of Combinatorial Theory*, 6 (1969), 20–32. The program is successful in arranging an unusually large number of interviews. Initially the work at a meeting was carried out locally on IBM punched card equipment. Then for a time it was done on the General Electric time sharing system as a courtesy. Since 1979 it has been done on the DEC 2060 in the Providence Office of the Society.

The Employment Register operated at a deficit for many years. This was recovered from general funds of the Society and the Association with a relatively small contribution from SIAM.

Mathematical Reviews

Rising tensions and unrest prior to World War II affected the world mathematical community. The reviewing journal *Zentralblatt für Mathematik und ihre Grenzgebiete* (*Zbl*) had been established in 1931 by the publishing house of Julius Springer, with an editorial committee of distinguished mathematicians from Europe and America. The editor was Otto Neugebauer, a professor at the University of Göttingen. His field is the history of mathematics and astronomy. He left Göttingen for a professorship at the University of Copenhagen in 1934, where he was in 1938 as events unfolded.

The Italian differential geometer T. Levi-Civita was one of the members of the original editorial committee. Although Mussolini initially distrusted the racism of Hitler's National Socialism, he later accepted anti-Semitism. As a consequence, Levi-Civita lost his professorship of rational mechanics at the University of Rome. His name was dropped from the Editorial Committee of *Zbl*. At about the same time, Russian reviewers were barred from *Zbl*. Springer, presumably under duress, required that no refugee Jews be allowed to act as reviewers and sought a written and binding guarantee of this policy from Neugebauer. At this point, Neugebauer resigned as editor. A number of the members of the Editorial Committee, including all of those from the United States, namely Richard Courant, J. D. Tamarkin, and Oswald Veblen, and others outside of Germany, also resigned. Between the persons not allowed to review and those who chose not to review in the face of the intrusion of political considerations into a scientific enterprise, the coverage of the journal weakened for lack of competent reviewers.

Before the end of 1938, informal consultations took place among mathematicians in the United States about how to save *Zbl* in such an emergency. At the Society meeting of December 1938 in Williamsburg, there was a meeting of the Council attended by about seventy-five additional persons by invitation. The issue of an abstracting journal in the United States was introduced by Secretary Richardson and was discussed by Thornton C. Fry and Warren Weaver prior to a general discussion. The outcome was the following resolution:

It is the opinion of this informal meeting that the American Mathematical Society should undertake the sponsorship of a new abstracting journal in mathematics for an initial period of five years, provided that a suitable mode of cooperation with foreign mathematicians, especially in Great Britain, can be effectively arranged and that adequate financial backing can be obtained.

In closed session, the Council, on a motion by Warren Weaver, requested the president "to appoint a committee, not to exceed seven in number, with power to carry out the preliminary studies and negotiations and, in case it is deemed wise, to proceed with the inauguration of a mathematical abstracting journal." The initial five year period, the necessity for adequate financial backing, and the appropriate cooperative arrangements were all imposed on the committee.

President R. L. Moore subsequently appointed the Committee, consisting of C. R. Adams, chairman, G. D. Birkhoff, A. B. Coble, T. C. Fry, Marston Morse, and G. T. Whyburn. Correspondence from the period shows that Tamarkin and Veblen were not considered "for diplomatic reasons," i.e., because they had so recently been on the *Zbl* editorial committee.

Initial funding for the committee was budgeted at $500 for travel and clerical expenses.

The report of the Committee on the Establishing in America of an Abstracting Journal in Mathematics at the Council meeting of 25 February 1939 included the news that the President of the Carnegie Corporation was recommending an appropriation of $60,000 "for the support of 'an international mathematical journal' over a preliminary period, 'to be sponsored by and *eventually supported by various mathematical societies.*' " Further, it is noted that Springer stated its intent to continue the *Zbl* "as an impartial abstract journal on the present high level" and its plan to send a representative to the United States to discuss questions related to the possible founding of another abstracting journal.

The Council approved a recommendation of an appropriation of $1000 annually for five years to support a mathematical abstracting journal sponsored by the Society, provided such a journal were inaugurated.

In April 1939, it was announced that O. Neugebauer and J. D. Tamarkin had been approached with a view toward their joint editorship. Neugebauer had already been offered an appointment at Brown University that he had accepted, was in the United States for about ten weeks beginning on 16 February 1939, and was available to advise the committee.

The name *Mathematical Reviews* (*MR*) for the proposed journal was in use by the end of May, at which time, the committee, having been authorized to decide whether to proceed with the journal did in fact so decide. The journal was to begin publication in January 1940 with coverage from 1 July 1939. The initial list price was $13 for one year, with a price of $6.50 to members of sponsoring societies. The objection of Professor W. C. Graustein that the price was too low was recorded.

There were alternatives and counter currents during the course of making the decision. One possibility that was considered was to buy *Zbl*. The representative from Springer, whose visit had been promised, was F. K. Schmidt. He indicated that the prestige of the German government required that *Zbl* continue.

Another potential problem was that a new journal in the United States would involve a break with German and Italian mathematicians, which might retard international good will in mathematics. This question was complicated by the fact that plans for the International Congress of 1940 were in progress and could be disrupted by such a break.

G. D. Birkhoff had a suggestion for a monthly classified listing of papers, with name and subject index, to be carried out at a clerical level. This would be accompanied by reviews of selected papers. A photocopying service of papers on request would be a substantial part of the venture. The whole was to be self-supporting. He was concerned with the facts that the proposed abstracting journal might not be funded, that it was unfriendly to undertake a journal similar to *Zbl*, and that reviewing would take a heavy toll of energy of young American mathematicians. His proposal did not prevail.

On the other hand, Richardson expressed the opinion that an abstracting journal would almost inevitably be undertaken in the United States. He asked what the Council would think if a Council committee, having been given power to proceed, decided not to do so and then an abstracting journal was started under other auspices. He took the position that the Society "must avoid any reference to political, religious or racial questions. We must under no circumstances put ourselves in a position of appearing to kill the *Zentralblatt*. We must study the question objectively and make up our minds as to what is best for mathematics as a whole and then proceed as cheerfully and efficiently as we can."

In June, the Council recommended acceptance of the gift from the Carnegie Corporation. The Trustees concurred. The Council further authorized the expenditure of at most $5000 by 1 September 1939 in furtherance of the founding of the journal. The Carnegie grant in fact became endowment for the journal.

The Rockefeller Foundation made a gift of $12,000, which was used for equipment, promotion, and to replace income that would later be derived from subscriptions. The expenditure of $5000 recommended by the Council was to be charged here.

The scale of the financial operation at the beginning is seen in the initial draft of an annual budget of $20,000. In fact, the expenditure for the fiscal year 1 December 1939 to 30 November 1940, which included the first eleven issues of *MR*, was $14,356.77.

In seeking support from other organizations, the Society naturally turned to the Mathematical Association of America. There was a difference of opinion within the leadership of the MAA over whether an expenditure for a venture primarily in support of research was appropriate. The issue arose just at the point that the MAA was conducting a general survey of the direction of its future activities. The chairman of the committee, R. E. Langer, appeared to be opposed both to the idea and to making a decision on it before the work of his committee was completed. The MAA noted that initial support seemed assured between the Carnegie Corporation gift and the proposed Society subvention and wondered whether spreading its support over an interval longer than five years might be helpful.

The Executive Committee for Mathematical Reviews, consisting of Oswald Veblen, chairman, T. C. Fry, and Warren Weaver, appointed Dr. Willy Feller to be technical assistant to the editors for a three year term effective 1 July 1939 and appointed Miss Evelyn Spencer to be a secretary for the editorial staff. Both salaries were charged to the Rockefeller subvention.

By deliberate decision of the committee, the names of the editors and the editorial assistant did not appear in the front matter of the new journal. The day to day work was done initially by Neugebauer and Tamarkin but then devolved on Feller. The name of Feller first appears, with the title Executive Editor in vol. 5 (1944). By this time, Oswald Veblen had joined Neugebauer and Tamarkin on the Editorial Committee, which was also listed for the first time.

Before the first issue of *MR* appeared, there were about 700 subscribers and 350 reviewers, with about 220 of these from the United States and Canada. The staff of reviewers was then regarded as almost complete. By 1943 there were 1332 subscribers.

As a promotional measure, the Society gave a premium consisting of a microfilm reading machine with initial three year subscriptions to the journal up until the middle of February 1942. This offer was subsidized by the Committee on Scientific Aids to Learning (itself supported by the Carnegie Foundation), which supplied machines of a simple design from the Spencer Lens Company. The machine had a retail value of $32.00. The initial annual subscription price of the journal to members was $13.00 but it was half that to members of sponsoring organizations. The initial stock of 400 machines was exhausted and more were made available.

The Society initially sold both microfilm copies and photocopies of the papers that it reviewed, except for books and copyrighted material. The price was two cents for an exposure (usually two pages) and fifteen cents for a photocopy (also two pages). This service ceased after 1947. The last five microfilm readers were offered for sale at a nominal price in 1956.

The first volume of *MR* in 1940 consisted of 400 pages with 2120 reviews. The size remained almost constant through 1945 but jumped abruptly to 621 pages in 1946. The size increased gradually through 1960 when it had reached 1600 pages. In 1961 there was a marked increase to 2548 pages. With the year 1962 came the change to two volumes per year and the number of pages per year continued to increase, passing 3000 in 1966 and 5000 in 1975. In 1977 and 1978 the number was a bit below 5000. In 1979 there were 7460 pages with 52809 reviews, corresponding to the fact that a backlog in reviewing had developed and was cleared. The journal then stabilized with more than 5000 pages and 34000–38000 reviews per year, numbers that gradually increased to keep pace with the volume of published research. In 1987 there were 7419 pages with 51848 reviews, as well as more than 3100 pages of indexes.

It was decided as early as 1943 not to copyright the journal. This policy was not reversed until the issue of September 1960.

The gathering of published mathematics during World War II presented difficulties and there was a decrease in publication. However, in November 1943 the Secretary noted that material from Russia was being obtained regularly and that material from Germany up to the beginning of 1943 had been secured. Some material was made available by the alien property custodian.

The gradual increase in journal size over almost fifty years represents chiefly the increase in the volume of published mathematics. The policy for selecting material for review has been nearly constant. The most serious issue may have been over applied mathematics. The policy set by Neugebauer was that a paper dealing with an application of mathematics should be reviewed on the basis of the new mathematics (if any) contained in it, without regard to the importance of the paper on other grounds.

The subscription price of the journal has increased with both the size of the journal and the general inflation. The size of the journal between 1940 and 1988 has increased by a factor of about 25. It is not completely clear what index of prices to use but it appears that a factor of about 8.4 between 1940 and 1988 is reasonable. A calculation yields a list price of $2730, compared with the existing list of $3275. On the other hand, the member's price calculated in the same way is $1365 whereas the actual price to members is $393.

The journal is supervised by an editorial committee of three persons (to be four after 1988), who are elected in an uncontested election to three year terms with the convention that one serve no more than two terms. The first elected editorial committee, chosen in 1941 for terms to begin in 1942 by virtue of a revision of the bylaws in 1941, consisted of O. Neugebauer, J. D. Tamarkin, and O. Veblen. Here are the persons who have served:

J. D. Tamarkin	1942–1945
O. Veblen	1942–1947
O. Neugebauer	1942–1948
M. H. Stone	1945–1949
W. Feller	1948–1953[1]
H. Whitney	1949–1954
E. Hille	1950–1955, 1961–1962
R. P. Boas	1954–1959, 1962–1963, 1964–1965[1]
J. L. Doob	1955–1960
W. S. Massey	1956–1961
J. V. Wehausen	1960–1965
H. A. Antosiewicz	1962–1967
P. R. Halmos	1964–1969
John Wermer	1966–1968
Richard S. Pierce	1968–1973
Frederick W. Gehring	1969–1974
O. Goldman	1970–1972
J. T. Schwartz	1973–1978
R. G. Bartle	1974–1976, 1983–1986[2]
Donald J. Lewis	1975–1978
W. J. LeVeque	1976–1977[2]
Paul T. Bateman	1977–1982
C. M. Pearcy	1978–1983
E. R. Berlekamp	1979–1981
M. Lowengrub	1982–1987
M. Hochster	1983–
L. D. Berkovitz	1986–
H. F. Weinberger	1988–

The editorial committee sets editorial policy for the journal, being concerned with such problems as breadth of coverage, the policy on cover-to-cover reviewing versus selective reviewing, critical versus factual reviews, the status of author abstracts as reviews, and matters of bibliographic style. One very important task of the committee is the selection of the executive editor, who on recommendation of the committee is employed by the Trustees.

---

[1] Boas served for Feller in February to August 1953.

[2] Bartle resigned during 1976 to become Executive Editor and the term of LeVeque began in midyear. He resigned again for the same purpose in 1986 and the term of Berkovitz began during 1986. LeVeque resigned in mid-1977 to become Executive Director

The operation of *MR* devolves on an executive editor and a professional staff. More recently, the staff has included a managing editor and an increasing number of associate editors. When the journal was started in 1939, with initial issue in January 1940, O. Neugebauer doubled as a member of the editorial committee and as editor. W. Feller was "technical assistant" from 1939 and appears to have been editor by the time the second volume appeared in 1941. The passage of authority from Neugebauer to Feller however is not clear. Feller signed himself managing editor in 1943 but this may have represented his capacity on the editorial committee. There are references to Feller as executive editor and, as already noted, he was so called in 1944 but the recommendation of Boas essentially names Boas as the first executive editor. The succession of executive editors was as follows:

1940	Otto Neugebauer
1941–1945	Willy Feller
1945–1950	R. P. Boas, Jr.
1950–1956	J. V. Wehausen
1956	S. H. Gould, G. Y. Rainich
1957–1962	S. H. Gould
(1961–1962	A. J. Lohwater, managing editor)
1962–1965	A. J. Lohwater
1965–1966	W. J. LeVeque
1966–1968	S. K. Berberian
1968–1971	R. J. Crittenden
1971–1976	J. Burlak
1976–1978	R. G. Bartle
(1976–1977	J. Burlak, senior editor)
1978–1986	John L. Selfridge
(1979–	W. B. Woolf, managing editor)
1986–	R. G. Bartle

As the succession of dates indicates, changes frequently took place in midyear, corresponding to the academic calendar.

Among the executive editors, Boas, Wehausen, Berberian, and Selfridge had their first substantial contact with the journal in that position. On the other hand, Lohwater was managing editor under Gould in 1961–1962 before becoming executive editor. LeVeque, Crittenden, and Burlak had all served as associate editors. Bartle moved from the editorial committee to the editorship twice during his terms on the committee as already noted.

The annual report from the editorial committee of *Mathematical Reviews*, then consisting of J. V. Wehausen, chairman, E. Hille, and W. S. Massey, for 1961 contains a passage that exhibits some changes in operation of the journal from the early days of close supervision by the Editorial Committee:

In October A. J. Lohwater was appointed to the *MR* editorial staff as 'Managing Editor.' He is responsible 'for matters concerning day-to-day operations of the office,' according to the Report of the Executive Director [Walker] and the Executive Editor [Gould], dated Nov. 16, 1961. 'Matters concerning the general policy of the journal, and the relationship of Mathematical Reviews with the outside mathematical world, continue to be the responsibility of the Executive Editor.' Essentially, what used to be the function of the Executive Editor has now become that of the Managing Editor, and what used to be the function of the Editorial Board has become that of the Executive Editor. There is now some question as to whether an Editorial Board is any longer necessary, or at least what its function should be. It had, for example, no prior knowledge of the above appointment or reassignment of duties.... The Board *did* meet together with the Providence staff on November 25th to discuss several questions of reviewing policy, coverage and indexing.

After the move to Ann Arbor noted below, closer supervision by the Editorial Committee returned. This was facilitated by a policy that one member of the committee was from the Michigan faculty.

After Lohwater, no managing editor apeared until W. B. Woolf was appointed in 1979 early in the tenure of Selfridge. Woolf served as acting executive editor during 1984–1985 while Selfridge was on study leave.

The first associate editor, and the only one in 1956, was G. Y. Rainich, who served only that year. From that time forward there were several associate editors at one time, the number reaching a maximum of fourteen in 1981. The staff of associate editors does not completely represent the intellectual support at that level inasmuch as consultants are also used for special languages and fields and sometimes simply for manpower.

The initial staff under Neugebauer and Feller consisted of two additional persons. By 1988, the staff had grown to more than 70.

Initially *MR* was a file-card office. Reviewers, assignment of papers to reviewers, the load carried by reviewers, and the progress of papers from receipt to publication was recorded on cards. A great deal of repetitive typing and proofreading was characteristic. The advent of the computer and of hardware and programs to handle large files was welcomed. The change was made over a period of several years until the entire operation was computerized by 1980. The locus changed in 1981 from a computer mainframe at the University of Michigan to the DEC in Providence, soon augmented by a second, accessed by a dedicated telephone line. The operation was then divided, with the clerical work of maintaining and updating the MR database done on computer terminals in Ann Arbor connected to the Providence DEC, the issue preparation done on a network of smaller machines in Ann Arbor, and only the keyboarding of review texts done in Providence, again on the DEC.

The first location of the editorial offices was Brown University. There is mention in December 1939 of a letter of thanks to Brown for furnishing housing and personnel, the latter presumably being Tamarkin and Neugebauer. The editorial office stayed at Brown University until 1951. At that time the general offices of the Society moved from New York to Providence to settle at 80 Waterman Street and *MR* moved to the same location. The two offices stayed together when the move to 190 Hope Street in 1956 and the move to Butler Health Center in 1962 took place. Over a period of time, there was friction. It is possible that conflict of personalities is a concomitant of the two offices, each managed by a strong personality. In any event, *MR* was moved to Ann Arbor on 7 June 1965. It was W. J. LeVeque to whom fell the task of executing the move. The University of Michigan was one of several places with the required qualifications of a good library, a large diversified staff (so that consultants would be available), and a receptive attitude.

On June 7, 1965 *MR* moved to Ann Arbor to space in a building at 416 Fourth Street, formerly a brewery, rented from the University of Michigan. In 1970 *MR* moved to space at 1315 Hill Street and in 1975 to an office building at 611 Church Street. In the spring of 1984, the Society bought the building at 416 Fourth Street and remodeled it to suit the needs of *MR*. The move from Church Street to Fourth Street was made on 9 November 1984.

From the beginning of the time of *MR* in Ann Arbor, the Society had a designated local representative in Michigan. So far as one could see this was a purely formal response to Michigan law. However, when the Society bought the property in Ann Arbor it became desirable that *MR* be a subsidiary corporation in Michigan so that favorable tax treatment could be obtained. Mathematical Reviews, Inc. was established on November 8, 1983.

The first executive editor beyond the nucleus of Tamarkin, Neugebauer, and Feller was Ralph Boas. I quote from his reminiscences:

> When I was invited to be a reviewer, I hesitated, but finally accepted. What I found was that by being a reviewer I learned many interesting things that I probably would not have seen otherwise. Since I generally wrote my reviews promptly, I soon began to receive a lot to review, especially after I began to review Russian papers. Feller was half-time editor and in 1945 he resigned from *MR* to be full time at Brown. I was at Harvard as a temporary replacement for faculty who were away doing war work; I was making $4000 at Harvard and asked AMS to pay me that much. The AMS thought this excessive, but since there was no other candidate, they gave in; but I got the reputation of being "grasping" with J.R. Kline. I spent several months in 1945 learning the job.

At that time, Neugebauer and Tamarkin were not spending much time at *MR*; Feller was half-time editor, the secretary was Janet Sachs; the manuscripts were sent out to a nonresident copy editor. When I arrived I discovered (to my surprise) that I was expected to take over the copy-editing as well as all the editorial work, which included assigning papers to reviewers, editing the reviews, reading proof, making the subject index, and tending to the correspondence. For the first few months, Feller and I sat at the same desk and discussed everything in detail. We felt under pressure to keep everything concise; once we succeeded in cutting a ten page review down to five pages. (Feller struck out "partial" from "partial derivative," saying, "What other kind is there?"). Neugebauer handled most of the correspondence with foreigners, partly because many of them were old-time acquaintances. It was through his friendship with Heinz Hopf that *MR* was able to keep up with the German literature throughout the war, via Switzerland. We could not keep up with either the Italian or the Japanese journals; when they eventually came in, there were no real surprises from Japan, and only one (Cesari's work on surface area) from Italy. Neugebauer made a point of always writing in English. When van der Waerden reproached him with not using his "Muttersprache," he wrote back, "It's not a question of my mother's language, but of my secretary's."

For the first few years, *MR* published 280 pages of reviews a year, about 7 reviews a page, or about 2000 reviews. The procedures were simple: on Monday we distributed the papers to reviewers; records were made and the journals went downstairs to the Photo Lab to be microfilmed (and copied, if the journals were borrowed from the library); on Friday the articles were mailed to the reviewers. The rest of the time I edited reviews, read proof, tended to correspondence, and went to the library to check references (in those days, every reference was verified from our files, or from the author index of the *Zentralblatt* (which Neugebauer brought with him on cards), or in the library).

Fortunately Lancaster Press was willing to work from legible handwritten copy, but some reviewers' hands were so bad that the reviews had to be copied. We never dared insist on reviewers' providing legible copy for fear of losing them, especially if they could read Russian or other, even stranger, languages. However, we did send uninformative reviews back to reviewers for clarification, not always successfully. Part of my job was to translate all titles that were not in English, French, German, or Italian.

One Monday after the war ended there were only three papers to distribute, until the mailman came with a large registered package of the war years' Roumanian journals. At that point the flood began. Presently Wehausen came back from Japan with the wartime issues of the Japanese journals, and we had to photocopy them, at a cost of, as I recall, $300, an expense about which the AMS was not enthusiastic. (It should be remembered that for a few years, *MR* had run at a modest profit.) It must have been about 1947 that Mark Kac walked into my apartment one evening, remarking "Say, Ralph, have you noticed how big the *Reviews* is getting?" Had I noticed, indeed. By 1950, *MR* was publishing 766 pages of reviews in a year, and of course it has accelerated ever since.

I calculated once that I read every review nine times; I doubt that there is such close supervision now. The selection of reviewers was a difficult part of the job, because although we had them indexed by field and languages, there were often no reviewers who had specified the field of a given paper and admitted to a knowledge of the relevant language. We had to be persuasive. Some reviewers were incredibly cooperative; some were the opposite.

We went, under Neugebauer's influence, to great lengths to achieve completeness and accuracy. However, we took a firmer line than is followed today about what should not be reviewed. Neugebauer's principle was that papers were to be reviewed for their mathematical content, no matter how significant they might be otherwise.

The relaxed atmosphere and the small staff described by Boas are a far cry from the journal of today, where there are an executive editor, a managing editor, an associate executive editor, about ten associate editors, and a total staff of about 75, not counting consultants who supplement the work of associate editors in exotic languages or out-of-the-way subject matter. Instead of the initial 2000 reviews per year, the number is now up to 51,000. One year of the journal, unbound, occupies one foot of shelf space and the index alone comes to another five inches.

The semi-autonomous state of *Mathematical Reviews* has been the object of repeated discussion. For many years the executive editor reported to the Trustees. Although the executive director was responsible for the total income and expenditures of the Society, he did not have firm control of the budget of *Mathematical Reviews*.

As early as 1948, the Committee on Reorganization recommended not only that the office of executive director be established but also that the editorial offices "be combined with the general offices of the Society, thus concentrating all editorial work requiring paid assistants under the direction of the Executive Director." The recommendation was qualified with remarks that the space shortage in the Society offices (then at Columbia) and the living arrangements of the executive editor (in Cambridge, MA) made the change undesirable at the moment.

When *Mathematical Reviews* and the general offices were in the same physical location, as they were in Providence from 1951 until 1965, there was internal conflict that was partly responsible for the separation when *Mathematical Reviews* moved to Ann Arbor. Repeated consideration of the possibility of bringing the two parts of the Society business together in one location, either Providence or a third location, foundered. Finally in 1987, when William H. Jaco was employed to be executive director beginning late in 1988, the Trustees endorsed the principle that the executive director is the chief executive officer of the Society, including *Mathematical Reviews*. This statement clarifies the wording of Article VI, Section 1, of the bylaws, where the scope of the executive director is defined. Moreover, the Trustees charged the executive director, the executive editor, the chairman of the board, and the associate treasurer (who pays special attention for the Trustees to the financial affairs of *Mathematical Reviews*), to effect the change in structure of the management during the first year of service of the next executive director.

A problem that had arisen in the course of previous discussion was the possibility of interference by the executive director in the editorial direction of *Mathematical Reviews* as opposed to his function in managing the business of *Mathematical Reviews* as part of the business of the Society. There is an ethical issue here because the Society publishes journals and books whose sale is the responsibility of the executive director but which are reviewed in *Mathematical Reviews*. Accordingly the new model of management was to be designed so as to protect the editorial integrity of *Mathematical Reviews*.

During its early years *Mathematical Reviews* operated at a surplus. Its subscriptions for the first five years were:

1940	1225
1941	1350
1942	1400
1943	1239
1944	1332

The income through 30 November 1944 (end of fiscal year) was \$83,047.12 in contributions (including \$60,000 from the Carnegie Corporation) and \$61,473.35 from subscriptions and other sources while the expenditures were \$67,584.38. The receipts and expenditures for 1944 were \$15,453.59 and \$12,770.02 respectively.

The rejuvenation of the *Zentralblatt* following World War II was a competitive influence that was closely watched. The report of the Editorial Committee of *Mathematical Reviews* in December 1949 notes that speed in reviewing was one of its aims, that it is ahead of the *Zentralblatt* on most papers, and that this is particularly true of Russian papers.

Mathematical Reviews has long offered a service for mathematicians of providing copies of papers from journals that are difficult to obtain. When this service was formalized in 1971 it was noted that no list of such journals would be established because the relatively inaccessible was different in different places but that the service did not apply to journals in nearby libraries.

The journal for many years was sold also in a one-sided printing for those who wished to clip it. The number of subscribers was always small (15 in 1948, 29 in 1949). More recently, as the bulk and price of the journal increased and the classification of papers was refined, subscriptions to sections of the classification were available.

Mathematical Reviews has spawned a variety of closely related services. There are cumulative author indexes for 1940–1959, 1960–1964, and 1965–1972. There are cumulative subject indexes for 1940–1958 and 1959–1972. Finally, there are combined author and subject indexes for 1973–1979 and 1980–1984. Earlier versions were produced by physical handling of file cards and transcription. With the increased computerization of the production of the initial reviews, the information for indexes is all in a data bank for almost automatic production of the indexes.

The database of *Mathematical Reviews* is an asset that has been exploited and made available through online vendors. Initially under the name Mathfile, the bibliographic information since 1973 and the full texts of reviews since July 1979 were made accessible. The material is distributed on the basis of royalties to the Society through three commercial services, Dialog Information Services, BRS (Bibliographic Retrieval Services) and ESA-IRS. The file of bibliographic information was later extended back to 1959. Data from *Current Mathematical Publications* was incorporated and replaced as the reviews appear. Other files, notably the current Index to Statistics and the Index to Statistics and Probability, 1902–1968, compiled by J. W. Tukey were added. The name was changed to MathSci in 1986 as an indication of the broader scope. On an ordinary screen or printer, the mathematical content is coded and takes some experience to read. The language of entries since 1985 is T$_E$X, so that with appropriate graphic display equipment or printer the coded mathematics can be made to appear in conventional form. There is an extensive user's guide to MathSci.

COMPUTERIZATION OF *MR*

Over a period of fifty years, the production of *Mathematical Reviews* changed. Whereas it was once a cardfile and coldtype operation it has gradually become completely computerized from the initial handling of papers to the preparation of camera copy. The following account has been slightly edited from a draft prepared for this volume by William B. Woolf.

At its inception in 1940, *MR* was produced following procedures developed at *Zbl* and brought to the USA by Otto Neugebauer. Information was typed and retyped, first onto several 3×5 cards for maintenance of several record chains, including the basic control sequence and the reviewer records, then onto reviewer forms, and again by the typesetter for the printed journal. It was typed yet again for the issue and for the volume indexes and, from 1973 on, for the annual indexes. When cumulative indexes were issued, the information was typed yet one more time. These repeated keyboardings, each with its potential for error independent of the others, made it extraordinarily difficult to maintain the reputation for accuracy which *Mathematical Reviews* had always sought. Over the years, efforts were made to find methods to reduce this multiple keyboarding. Sometime in the late 50s, the "Ditto" machine was introduced into the process: a Ditto master was typed of each reviewable item, and both 3×5 cards and review forms were printed from the Ditto master. This eliminated several keyboardings at the front end of the production pipeline, but left the process prone to independent error at the issue, issue index, annual index and cumulative index stage. In fact, as late as 1978 it was estimated that bibliographic information about reviewed items was being keyboarded as many as 10 times.

The first major effort by the Society to computerize (the word then was "mechanize") the handling of bibliographic information appears to have been initiated in the late 1960s. Under an NSF grant, concerted efforts were undertaken to utilize punched tape input to Photon type-setting machines for the production of indexes. Photon machines in another version were used in the Providence office for the production of the annual *Indexes to Mathematical Publications* (IMP), Volumes 1–4, which covered the literature in the Mathematical Offprint Service, already described, for the periods July–December 1970, 1971, 1972, and 1973, respectively. When the NSF ended funding of MOS, it was replaced by the less elaborate Mathematical Title Service and the Indexes to Mathematical Publications. That service ended in 1973, and Volumes 5–9 of IMP were annual indexes of *Mathematical Reviews*. The first four of these indexes were prepared using IBM Selectric typewriters with many different typing elements providing both font selection (Roman, italic, bold, Greek, Cyrillic) and mathematical symbols. Finally, the index for the 1977 issue year of *MR* was typeset using the STI typesetting system (see below) but still requiring a complete re-keyboarding of the information.

Finally, in the late 1970s, the production procedures at *MR* were effectively computerized. The first step succeeded in computerizing the reviewer files, including procedures for keeping track of the papers assigned to reviewers and producing reminder letters intended to encourage prompt return of reviews from sometimes laggard reviewers. Immediately on the heels of this project came the effort to computerize completely the management of bibliographic information concerning items covered in *Mathematical Reviews* and *Current Mathematical Publications*. The goal was to ensure that one initial keyboarding would suffice for all production steps, from the initial record keeping, through the reviewing stage, and finally including the typesetting of issues, issue indexes, volume indexes, annual indexes, cumulative indexes, and specialized review volumes.

The project was carried out in 1978–1979 under the leadership of Executive Editor J. L. Selfridge using a program called SPIRES developed at Stanford University for the management of bibliographic information. SPIRES (Stanford Public Information Retrieval System) was available on the University of Michigan computer system.

During the same period the British typesetting house William Clowes & Sons Ltd., which had typeset *MR* for years, reorganized and, in the process, raised the price for setting *MR* to such heights that alternatives were sought. The decision was made to typeset *MR* in the STI (Science Typographers, Inc.) system, with the review text being input in the Providence office of the Society while the bibliographic information would be input and maintained in the *MR* database created in Ann Arbor. Thus when a galley or an issue was to be printed, a program on the Providence computer took a bibliographic file from the *MR* database and a file of review texts input in the Providence office, merged them, and ran them through the STI program to create a file which drove a Harris typesetting machine (first in Washington, D.C., and later at the STI headquarters on Long Island). [The result was a system which provided typeset output for *CMP*, *MR* and the several indexes, but output of proof-quality for internal record-keeping or for galleys was in coded form, requiring staff and reviewers to learn (or at least tolerate) the special typesetting codes imbedded in the bibliographic files; in fact, galleys were printed by the Harris typesetting machine in order to provide editors with a final-quality version of the output for proofreading.]

In Spring, 1981, work began on the migration of the *MR* database created in Ann Arbor. Thus, when a galley or an issue was to be printed, a program on the Providence computer took a bibliographic file from the *MR* database from the SPIRES system on the University of Michigan Amdahl computer to the SEED database management system on the DEC 2060 machine in the Providence office. That project continued over the next two years, involving the efforts of staff members of the Providence Computer Services Division (CSD) as well as staff members of *MR* in a detailed analysis of the production of *MR* and the design of the database system. The latter, with enhancements has persisted.

Over the first half of the decade of the 1980s, staff members of both offices grew increasingly familiar with TEX, the typesetting language especially designed by Donald E. Knuth of Stanford University for the setting of mathematics. During 1984, the staff at *MR* undertook the necessary programming steps to move *MR* from the STI typesetting system to TEX. Effective with the January, 1985, issues of both *MR* and *CMP* were typeset in TEX. This meant that the initial keyboarding of bibliographic information could be used without emendation to provide both proof quality output for all pre-publication uses (internal record-keeping documents, reviewer forms, galleys, etc., using either dot-matrix impact printers or, later, laser printers) and final camera-ready output for the several printed versions of the information (*CMP*, *MR*, *MR Sections*, review volumes, issue indexes, annual indexes, specialized indexes), using phototypesetting machines such as those of Alphatype or Autologic.

In 1983, the Digital Equipment Corporation (DEC) announced that it would cease development of the DECsystem 20 (the Society owned two DEC 2060s, and had been expecting to be able to migrate to the rumored 2080 system). Plans were immediately laid in each office to examine alternative hardware choices and to lay out development programs leading to independence from the DEC 20s sometime in 1990.

MR examined the microcomputer work-station environment as well as the VAX family of super-minicomputers (which CSD had selected as the environment in which to develop the AMS database, AMSDB, used for all membership and business activities of the Society). A decision was made to move the typesetting (TEXing) aspect of *MR*'s computing from the DEC 2065 (the improved version of the DEC 2060 then in use) to Apollo work stations, and to develop the new version of the *MR* database on the Providence VAX machines, utilizing the database management system INGRES which had been chosen by CSD for the AMSDB. Work began in earnest in the Spring of 1987, and is expected to be complete late in 1989 or in 1990. It is being carried out in a way to make it relatively straightforward to migrate to the work station environment when machine power and software developments are far enough along. There is the possibility that the system will be developed in such a way as to allow on-line editing of manuscripts by editors and proofreaders, utilizing an interactive TEX program developed by a small Ann Arbor Company, ArborText, whose founders were all active in the first successful computerization of *MR* in the late 1970s.

MATHEMATICAL REVIEWS MERGER

It is intended to give a brief history of the *Zentralblatt* as it is related to *Mathematical Reviews* as background to discussion of a possible merger of the two which, in the end, did not take place.

The journal with the formal name *Zentralblatt für Mathematik und ihre Grenzgebiete* was founded in 1931. Throughout its existence it has been published as a commercial venture by Springer-Verlag (before 1942 the name was Julius Springer), one of many such journals from the same publisher. The initial editor, who served through 1938, was Otto Neugebauer.

Zbl almost expired during WWII. It was recognition of the failings of the journal, such as inadequate coverage of fields because of a shortage of reviewers, some of it politically induced, that led to the founding of *MR* by the AMS, as has been noted.

Beginning with volume 23 in 1941, the cooperation (einverständnis) of the Deutsche Mathematiker-Vereinigung in the publication of *Zbl* is listed.

The division of Germany is reflected beginning with volume 126 of 1966–1967, when the journal is listed as edited (herausgegeben) by the Deutscher Akademie der Wissenschaften zu Berlin and the Heidelberger Akademie der Wissenschaften. With volume 233 of 1971, the former name was replaced by the Akademie der Wissenschaften der DDR.

In 1971 the AMS (MOS) classification scheme was adopted by *Zbl* and in 1974 the name *Mathematical Abstracts* was added to the original name.

Beginning with volume 352 of 1978, the divided editorial management was gone, only the Heidelberg component remaining. Volume 382 of 1979 shows a new participant, the Fachinformationszentrum Energie Physik Mathematik GmbH (FIZ), which is listed alone in that single volume but is listed along with the Heidelberger Akademie thereafter.

Beginning about 1978, the *Zbl* has a tripartite management. Springer, as publisher, composes and distributes *Zbl* and funds the editorial offices in Berlin. The setting of editorial policy rests with the Heidelberger Akademie but FIZ is deeply concerned with the production of the journal because of financial responsibility for the editorial staff and its interest in (and funding of) developing and marketing the associated database.

The issue of merger of *MR* and *Zbl* has been raised several times. It is superficially appealing to say that one journal instead of two would take half the work at half the cost. During the time that *Zbl* was associated with both sides of the divided Germany, the exploration was begun and then terminated for two reasons. The complexity of dealing with a publisher and two academies on one side was clearly a disadvantage. Moreover, the direction of the conversations shifted in that Springer appeared to interpret the suggestion of merger as a veiled offer by the AMS to sell *MR*.

Cooperation between *MR* and *Zbl* has occurred in several areas. First, each has given to the other the right to reprint reviews. Second, the two journals have jointly devised and established the subject classification scheme now used by both. Moreover, the two journals have profited from the exchange visits in which staff members of each have observed the production methods of the other.

In 1982 the question of merger was raised again and was very seriously considered. It grew from a visit during the week of 4–10 October 1982 by Professor B. Wegner, editor in chief of *Zbl*, and Mr. O. Ninneman, director of the Scientific Staff, to the office of *MR*. The two visitors had extensive discussions with J. L. Selfridge, executive editor of *MR*, W. B. Woolf, managing editor, and members or past members of the *MR* Editorial Committee and associate editors. The group drafted a policy statement that reads as follows:

> It is proposed that *Mathematical Reviews* and *Zentralblatt für Mathematik* agree in principle that the jobs they perform should be restructured to eliminate duplication and that their responsibilities include the preparation of subject overviews and further development of a joint mathematical database and thesaurus. This would make efficient and effective use of their resources and better serve the mathematical community of the world.
>
> Conversations leading to the implementation of this principle should continue during 1982–1983 between representatives of the organizations involved.

With this statement before it, the Executive Committee and Board of Trustees (1) made a recommendation to the Council that resulted in a Council resolution to "approve the principle of cooperation between and possible merger of *Mathematical Reviews* and *Zentralblatt*."

Representatives of the interests in *Zbl* had adopted a similar statement in November 1982 following the October meeting.

With the Council resolution as its foundation, a negotiating team acted for the Society. It consisted of Alex Rosenberg, chairman, William J. LeVeque, Franklin P. Peterson, John L. Selfridge, and William B. Woolf. Rosenberg was a Trustee. When he resigned as Trustee toward the end of 1983, his place as chairman was taken by Steve Armentrout. When Selfridge relinquished the post of executive editor, his place on the team was taken by his successer, Robert G. Bartle. The actual negotiators included others at various times, notably Melvin Hochster, a member of the *MR* Editorial Comittee, Salvatore Trofi, fiscal manager, and Jane E. Kister, associate executive editor.

The chairman of the group representing *Zentralblatt* was Professor Dieter Puppe (Heidelberg Academy of Science). Other members were W. Rittberger (FIZ), E. Schultze (FIZ), B. Wegner (Editor-in-Chief of *Zbl*), O. Ninnemann (director of the Scientific Staff of *Zbl*), and W. Beiglböck (Springer-Verlag). FIZ is the West German government agency already mentioned, which has been the vehicle for a government subsidy of *Zbl*.

Negotiations were slow to begin. In February 1983, Peterson, Rosenberg, Selfridge, and Woolf met in Karlsruhe with representatives of FIZ, the Akademie, and the Berlin office and Rosenberg, Selfridge, and Woolf visited the Berlin office for several days. In the Summer of 1983, Woolf spent the month of July in Germany, with three weeks in Berlin and two visits to Karlsruhe and to Heidelberg. There were discussions and exchanges of information such as income and statistical data on subscriptions in the Summer and Fall of 1985. There was a meeting of the two teams in the United States in Providence in October 1985 and another meeting in Schwarzenberg, West Germany in April 1986.

The negotiations were not intended to effect merger but to approach a position of agreement or near agreement that could then be examined by the principals. An American position and a German position were achieved, reasonably close together, and concerned with the limited aspect of a printed journal.

The state of the negotiations following the two sessions is summarized in the following extract from a report prepared by Armentrout for the Council of 20 January 1987:

2. Axioms

The negotiations have focused on the idea that the two sides would cooperatively produce a database from which a printed review journal and possibly other products would be produced. Both the present offices would be maintained, possibly at reduced staff levels.

Since fall, 1985, the negotiations have assumed certain axioms. Among these are the following:

I. Both sides would contribute equally to the merged journal (in particular, equality in number of items processed by each office, input effort, and quality of work).

II. Each side would take care of its own expenses.

III. Division of income from the merged printed journal would be based on a fixed geographical division of the world (possibly excluding some parts of Asia) with each side the exclusive distributor of the printed merged journal in its region and with that side taking the resulting subscription income. (There would perhaps be a transition period.)

IV. There would be essentially a uniform price worldwide. Discounts by a side would be allowed, at that side's expense.

The two teams' proposals make Axiom III above specific.

3. Present proposals

The merged reviewing journal would be called *Mathematics Abstracts and Reviews (MAR)*.

a. The proposals are in substantial agreement concerning the division of income for the printed journal *MAR*. The resulting division of the market (using 1985 subscription figures) yields a ratio of $5 : 2$ in favor of the American side.

b. Both teams proposed to merge the online services but there was no agreement on the division of income from such a service.

c. Other present products, such as *Current Mathematical Publications, Mathematical Reviews Sections*, etc., were discussed but no specific proposals concerning them were made.

d. There was substantial discussion of electronic communication (online, CD-ROM, etc.) but no specific proposals.

4. Scientific nature of *MAR*

The staff discussions which occurred with the two negotiating sessions were in part directed to a specification of the merged journal *MAR*. While some of these issues have been discussed by the negotiating teams, they remain the subject of staff deliberation. However, there seems to be general agreement in several areas; these are, of course, subject to further negotiation.

The staffs envisage a division of labor in which the journals and books of the world would be divided evenly (possibly on a geographic basis) between the Berlin and Ann Arbor offices. Each staff would continue its current treatment practices (autorreferat, author summaries, independent reviews, editorial treatments, etc.) pending explicit joint decisions to do otherwise. Similarly, each office would continue current practices of copyediting, referencing and proofreading; both would utilize recently developed programs and procedures for verification of author identities. Initially the scope of *MAR* would be roughly the union of the (roughly the same) scopes of *MR* and *Zbl*.

Decision as to the form and extent of indexing would be referred for discussion to the existing editorial committees. The two offices have undertaken a revision of the 1980 Mathematics Subject Classification scheme, scheduled for implementation in *CMP* in 1989 and in *MR* and *Zbl* (or in *MAR*) in 1990.

The staffs have proposed an 8.5×11 inch two-column format in a typeface similar to that now in use in both journals; they propose a journal appearing 17 times per year, plus an annual index.

The proposed merger would reduce the total effort by mathematicians in preparing reviews. This is in contrast to the value of having two separately prepared reviews for many papers. Those libraries that subscribe to both *MR* and *Zbl* would be saved the cost of a subscription. The conduct of the operation from two offices would entail some inefficiency. It would probably require a layer of supervision, as by a steering committee, and frequent interoffice visits.

Estimates were made of the possible net effect of the proposed merger on the finances of the AMS. There were enough imponderables that this was difficult. With a list price for *MAR* at least that of the predecessors (which were approximately equally priced in 1987 following a small increase for *MR* and a large one for *Zbl*) and with no discounts of any kind (in particular no institutional member discount) estimates of the annual net effect for AMS varied from +\$200,000 to −\$200,000. The effect of institutional member discount is to subtract \$235,000 from these bounds. That is, the proposed merger could well have lost money for AMS.

The ECBT had watched the negotiation with interest and had reviewed interim reports several times. Initially they had gone on record as favoring the merger although the votes were divided and, by some members who were in favor, not enthusiastic. In the final analysis, a vote among the Trustees to support the negotiating position of the American team did not pass and the Executive Committee recommended that efforts toward merger be terminated. The Council of 20 January 1987 followed the advice of the Executive Committee in recommending to the Trustees that negotiations be terminated.

Conferences

The pattern of meetings supported by the Society changed with time. Initially the only meetings were the Annual and Summer Meetings and the regional meetings, all of general mathematical scope and with programs of invited addresses and contributed papers. Meetings on specialized topics were not part of the plan though invited lectures clustered in a field were from time to time part of the general programs. Beginning with the Symposia in Applied Mathematics just after WWII, this state of affairs changed. There were introduced the Summer Institutes in pure mathematics, the Summer Seminars in applied mathematics, the Symposia on Mathematical Questions in Biology, and the Summer Research Conferences. Each was accompanied with appropriate opportunities for publication.

Symposia in Applied Mathematics

When World War II became imminent, the Society supported interest in applied mathematics through a Committee on Preparing Addresses in Applied Mathematics consisting of Richard Courant, chairman, R. M. Foster, and Harold Hotelling. The committee helped prepare three symposia in applied mathematics in 1941. One was a half day symposium in New York on 21 February 1941, which consisted of an address by J. J. Stoker on Mathematical Problems Connected with the Bending and Buckling of Elastic Plates and another by W. A. Shewhart on Mathematical Statistics in Mass Production. The second was a half day in Washington on 3 May 1941 devoted to a Symposium on the Rayleigh-Ritz Method and Its Applications with lectures by a mathematician, an engineer, and a physicist, namely Richard Courant, J. P. Den Hartog, and H. M. James. The third was a half day on 30 December 1941 with a lecture by L. V. Bewley, an electrical engineer, on The Mathematical Theory of Traveling Waves and one by I. S. Sokolnikoff on Some New Methods of Solution of Two-dimensional Problems in Elasticity. The committee recommended a more permanent program than these ad hoc arrangements.

The interest of many mathematicians in applied mathematics and in the applications of mathematics was stimulated by the work in such fields, into which they were thrust in World War II. The time was ripe for consideration of the role of the Society when the war was over.

A Special Committee on Applied Mathematics, with J. L. Synge as chairman, and Ruel V. Churchill, Richard Courant, Griffith Conrad Evans, W. T. Martin, John von Neumann, and J. W. Tukey as members, made several recommendations to the Council of 26 and 27 December 1946. These were adopted by the Council and included the following:

> That a Committee on Applied mathematics be appointed to organize programs in applied mathematics and to make recommendations to the Council in all matters pertaining to the interests of applied mathematics.
>
> That the Society establish an annual symposium on applied mathematics, the Committee on the Role of the Society in Mathematical Publication being requested to study the possibility of the publication of the proceedings of the symposium in book form.

The Committee on Applied Mathematics was duly appointed by President Hille and consisted of J. L. Synge, chairman, Richard Courant, G. C. Evans, John von Neumann, William Prager, and Warren Weaver.

It should be noted that at this time there was considerable debate over the possibility of establishing a division of applied mathematics within the Society. The Special Committee on Applied Mathematics was initially opposed to the idea in the Fall of 1946. In fact von Neumann alone voted for the idea at first but changed his vote. However an informal poll of a number of interested mathematicians showed substantial support as well as opposition, so the committee debated the matter. In a report dated 6 March 1947, the committee did take the position that no separate division should be established at that time but that the issue should be further considered. It is not clear that it was in fact considered.

In August 1947, the first of a continuing series of Symposia in Applied Mathematics was held. These were an almost annual event. The pattern from the beginning consisted of invited lectures and contributed papers on a designated topic. The first three were in the Summer but the standard time changed to the Spring with the seventh in 1955. They were thenceforth usually held contiguous to or overlapping a regional meeting, almost invariably in the East.

Beginning with the symposium of 1967, the Committee on Applied Mathematics became a joint committee with the Society for Industrial and Applied Mathematics. The last of the jointly sponsored symposia in 1983 no longer suited the needs or the task of SIAM, which withdrew sponsorship of the symposia though not from other work of the selection committee.

The first symposium, on Nonlinear Problems in Mechanics of Continua, was cosponsored by the American Institute of Physics, the American Society of Mechanical Engineers, and the Institute of Aeronautical Sciences. Such multiple sponsorship continued only through the third symposium though there was frequently a second sponsor.

Financial support for these conferences came from many government agencies including the Office of Naval Research, the Army Research Office (Durham), the Air Force Office of Scientific Research, or occasionally the Institute for Defense Analyses (Princeton).

The roles of the committees associated with these symposia have varied and the records that cover the roles are scanty. The pattern that ultimately evolved was that proposals were solicited from individuals or groups for a topic together with a possible organizing committee and speakers. The Committee on Applied Mathematics selected one of these proposals and authorized an organizing committee to produce a firm list of invited speakers. Personnel in the Providence office incorporated the material supplied by the organizing committee into a grant proposal to obtain funds to pay travel costs of speakers and perhaps of some participants, together with incidental costs. In earlier days there were manuscript fees as well. The plan to publish proceedings was part of the grant proposal.

In practice it was sometimes necessary for the Committee on Applied Mathematics to develop a topic and importune an organizing committee to take it over. Sometimes a member of the Committee on Applied Mathematics turned up as chairman of the organizing committee and editor. The personnel of some early committees suggests that they functioned as local arrangements committees rather than or as well as committees on invitations.

There was a separate Editorial Committee on Applied Mathematics Proceedings for a time but later the editor of a particular set of proceedings was designated by the Committee on Applied Mathematics. The matters concerning the reprinting of earlier volumes of proceedings devolved on the Committee on Applied Mathematics.

Proceedings of the first two symposia were published by the Society. Volumes 3 through 8 were published by the McGraw Hill Book Company in an arrangement whereby the Society retained the title but received no royalty. Then in 1957 the Society undertook publication itself. The Society purchased the remaining copies of volumes 3 through 8 and now lists them with the later volumes in the series *Proceedings of Symposia in Applied Mathematics* (PSAM, sometimes familiarly known as "PSALMS"). When SIAM became a cosponsor, this use of PSAM ended though other material appeared subsequently in the series. The jointly sponsored symposia appeared in a new series AMS-SIAM Proceedings.

Here is a list of the symposia in applied mathematics from the first phase:

Nonlinear Problems in Mechanics of Continua
Brown University, 2–4 August 1947
The program had already been planned by Brown University and was approved. There was no organizing committee.
Editor: E. Reissner
Publication: PSAM vol. 1.

Electromagnetic Theory
Massachusetts Institute of Technology, 29–31 July 1948
Committee on Program and Arrangements: W. T. Martin, chairman, T. R. Hollcroft, J. A. Stratton, J. L. Synge
Editor: A. H. Taub
Publication: PSAM vol. 2.

Elasticity
University of Michigan, 14–16 June 1949
Committee on Program and Arrangements: W. Bartky, R. V. Churchill, E. L. Ericksen, G. E. Hay, D. L. Holl, J. W. T. Youngs
Editor: R. V. Churchill
Publication: PSAM vol. 3.

Fluid Dynamics
University of Maryland, 22–23 June 1951
Arrangements Committee: M. H. Martin, chairman
Editor: M. H. Martin
Publication: PSAM vol. 4.

Wave Motion and Vibration Theory
Carnegie Institute of Technology, 16–17 June 1952
Arrangements Committee: A. E. Heins, chairman
Editor, A. E. Heins
Publication: PSAM vol. 5.

Numerical Analysis
Santa Monica City College, 26–28 August 1953
Arrangements Committee: D. H. Lehmer, chairman, H. F. Bohnenblust, John Hamilton Curtiss, J. W. Green, G. Pólya, John Todd
Editor: John Hamilton Curtiss
Publication: PSAM vol. 6.

Mathematical Probability and its Applications
Polytechnic Institute of Brooklyn, 14–15 April 1955

Organizing Committee: H. W. Bode, chairman, R. Courant, D. H. Lehmer,
T. Y. Thomas
Editor: L. A. Mac Coll
Published under title *Applied Probability*: PSAM vol. 7.

Calculus of Variations and its Applications
University of Chicago, 12–13 April 1956
Organizing Committee: William Prager, chairman, L. M. Graves, Norman
Levinson, Marston Morse, J. J. Stoker
Editor: L. M. Graves
Publication: PSAM vol. 8.

Orbit Theory
New York University, 4–5 April 1957
Organizing Committee: J. B. Rosser, chairman, G. Birkhoff, W. J. Eckert,
Philip Hartman, P. Herget, H. E. Newell
Editors: G. Birkhoff, and R. E. Langer
Publication: PSAM vol. 9.

Combinatorial Designs and Analysis
Columbia University, 24–26 April 1958
Organizing Committee: Marshall Hall, Jr., chairman, R. Bellman, G. E.
Kimball, S. A. Schelkunoff, C. B. Tompkins
Editors: R. Bellman and Marshall Hall, Jr.
Published under title *Combinatorial Analysis*: PSAM vol. 10.

Nuclear Reactor Theory
New York, 23–24 April 1959
Organizing Committee: Ernest P. Wigner, chairman, G. Birkhoff, H. L.
Garabedian, S. M. Ulam, J. E. Wilkins
Editors: G. Birkhoff and Ernest P. Wigner
Publication: PSAM vol. 11.

Structure of Language and Its Mathematical Aspects
New York, 14–15 April 1960
Organizing Committee: R. Jakobson, chairman, E. S. Klima, Secretary
Editor: R. Jakobson
Publication PSAM vol. 12.

Hydrodynamic Instability and Related Problems
New York, 14–15 April 1960
Organizing Committee: Garrett Birkhoff, chairman, R. Bellman, E. Hopf,
C. C. Lin, B. MacMillan
Editors: R. Bellman, G. Birkhoff, and C. C. Lin
Published under title *Hydrodynamic Instability*: PSAM vol. 13.

Mathematical Problems in the Biological Sciences

Mathematical Problems in the Biological Sciences
New York, 5–7 April 1961
Organizing Committee: S. M. Ulam, chairman, Richard E. Bellman,
Anthony Bartholomay, J. Jacquez, T. Puck, C. E. Shannon
Editor: R. Bellman.
Publication: PSAM vol. 14.

Experimental Arithmetic
Chicago, 12–14 April 1962
Organizing Committee: N. C. Metropolis, chairman, Marshall Hall, Jr.,
P. Henrici, M. Kac, R. D. Richtmyer, A. H. Taub

Interactions between Mathematics Research and High Speed Computing
Atlantic City, 16–17 April 1962
Organizing Committee: J. Todd, chairman, G. E. Forsythe, P. D. Lax,
D. H. Lehmer, H. H. Goldstine, C. B. Tompkins, D. M. Young
The preceding two symposia were published jointly under the title
Experimental Arithmetic, High Speed Computing, and Mathematics:
PSAM vol.15.
Editors: N. C. Metropolis, A. H. Taub, J. Todd, and C. B. Tompkins

Stochastic Processes in Mathematical Physics and Engineering
New York, 30 April–2 May 1963
Organizing Committee: R. Bellman, A. T. Bharucha-Reid, M. Kac,
J. M. Richardson, Joseph B. Keller, L. Zadeh, D. Slepian
Editor: R. Bellman
Publication: PSAM vol. 16.

Applications of Nonlinear Partial Differential Equations in Mathematical
Physics
New York, 21–23 April 1964
Organizing Committee: R. Finn, chairman, C. B. Morrey, W. Noll,
J. B. Serrin, A. H. Taub
Editor: R. Finn
Publication: PSAM vol. 17.

Magneto-fluid and Plasma Dynamics
New York, 13–15 April 1965
Organizing Committee: H. Grad, chairman, A. Lenard, M. N. Rosenbluth,
William R. Sears, H. Weitzner
Editor: H. Grad
Publication: PSAM vol. 18.

Mathematics Aspects of Computer Science
New York, 5–7 April 1966
Organizing Committee: J. T. Schwartz, chairman, M. Davis,

H. H. Goldstine, D. H. Lehmer, J. Todd, Herbert S. Wilf,
C. C. Elgot, S. Gorn, H. Huskey, A. G. Oettinger
Editor: J. T. Schwartz
Publication: PSAM vol. 19.

The following constituted the second phase jointly with SIAM:

Transport Theory
New York, 5–7 April 1967
Organizing Committee: G. Birkhoff, chairman, R. E. Bellman, H. Grad,
M. Krook, J. E. Moyal, T. W. Mullikin
Editors: I. K. Abu-Shumays, R. E. Bellman, G. Birkhoff
Publication: SIAM-AMS Proc. vol. 1.

Numerical Solution of Field Problems in Continuum Physics
Durham, NC, 5–6 April 1968
Organizing Committee: G. Birkhoff, chairman, Jim Douglas, Jr.,
R. S. Varga, Calvin Wilcox, Signey Fernbach
Liaison Representatives: Francis G. Dressel, A. S. Galbraith, Gene B.
Parrish
Editors: G. Birkhoff, R. S. Varga
Publication: SIAM-AMS Proc. vol. 2.

Mathematical Aspects of Electrical Network Theory, 2–3 April 1969
Organizing Committee: Herbert S. Wilf, chairman, W. A. Blackwell, Frank
Branin, Robert Brayton, Frank Harary
Editors: F. Harary, H. S. Wilf
Published as *Mathematical Aspects of Network Analysis*: SIAM-AMS Proc.
vol. 3.

Computers in Algebra and Number Theory, New York, 25–26 March 1970
Organizing Committee: Garrett Birkhoff, chairman, Charles C. Sims,
D. G. Higman
Editors: G. Birkhoff, M. Hall, Jr.
Publication: SIAM-AMS Proc. vol. 4.

Mathematical Aspects of Statistical Mechanics, New York, 7–8 April 1971
Organizing Committee: Mark Kac, O. E. Lansford III, James C. T. Pool,
chairman, Robert T. Powers, Seymour Sherman
Editor: J. C. T. Pool
Publication: SIAM-AMS Proc. vol. 5.

Stochastic Differential Equations, New York, 29–30 March 1972
Organizing Committee: Joseph B. Keller and Henry P. McKean, cochair-
men
Editors: J. B. Keller, H. P. McKean
Publication: SIAM-AMS Proc. vol. 6.

Complexity of Real Computational Processes, New York, 18–19 April 1973
Organizing Committee: Richard M. Karp, chairman, Stephen Cook, John Hopcroft, Shmuel Winograd
Editor: R. M. Karp
Published as Complexity of Computation
Publication: SIAM-AMS Proc. vol. 7.

Mathematical Aspects of Chemical and Biochemical Problems and Quantum Chemistry, New York, 10–11 April 1974
Organizing Committee: Donald S. Cohen, chairman, Hirsh G. Cohen, Julian D. Cole, George R. Bavalas, Aron Kupperman
Editor: D. S. Cohen
Publication: SIAM-AMS Proc. vol. 8.

Nonlinear Programming
New York, 23–24 March 1975
Organizing Committee: Richard W. Cottle, chairman, Carleton E. Lemke, Stephen M. Robinson, J. Ben Rosen
Editor: R. W. Cottle
Publication: SIAM-AMS Proc. vol. 9.

Asymptotic Methods and Singular Perturbations
New York, 11–12 April 1976
Organizing Committee: Robert E. O'Malley, Jr., chairman, Donald S. Cohen, Joseph B. Keller, M. D. Van Dyke
Editor: R. E. O'Malley, Jr.
Publication: SIAM-AMS Proc. vol. 10.

Computational Fluid Dynamics
New York, 16–17 April 1977
Organizing Committee: Alexandre Chorin, Herbert B. Keller, chairman, Peter D. Lax, R. W. MacCormick
Editor: Herbert B. Keller
Publication: SIAM-AMS Proc. vol. 11.

Mathematical Problems in Fracture Mechanics
New York, 28–29 March 1978
Organizing Committee: Keiti Aki, Robert Burridge, chairman, James K. Knowles, James R. Rice
Editor: Robert Burridge
Published as *Fracture Mechanics*
Publication: SIAM-AMS Proc. vol. 12.

Mathematical Psychology and Psychophysiology
Philadelphia, 15–16 April 1980
Organizing Committee: W. K. Estes, Stephen Grossberg, chairman,

R. Duncan Luce, M. Frank Norman, H. Simon, George Sperling
Editor: Stephen Grossberg
Publication: SIAM-AMS Proc. vol. 13.

Inverse Problems
New York, 12–13 April 1983
Organizing Committee: Robert Burridge, Joseph B. Keller, Robert B.
Marr, David W. McLaughlin, chairman, C. Ray Smith
Editor: D. W. McLaughlin
Publication: SIAM-AMS Proc. vol. 14.

Summer Institutes

An ad hoc Committee on a Summer Mathematics Institute, consisting of A. A. Albert, Chairman, S. Eilenberg, S. Mac Lane, D. C. Spencer, and O. Zariski, presented a report that the Council approved on 11 August 1952. The committee noted the profound influence of the Institute for Advanced Study but the defect that it did not allow for the gathering together of a group of mathematicians all interested in a single subject. The committee proposed "that a Summer Institute without a building or permanent staff be organized to provide such gatherings of mathematicians active in a particular subject." Support by the National Science Foundation was proposed as well.

A standing Committee on Summer Institutes was then appointed, consisting of A. A. Albert, chairman, and S. Bochner for three years, H. P. Robertson and H. Whitney for two years, and N. Jacobson and O. Zariski for one year. The committee immediately established the first institute for the Summer of 1953.

Several principles were established in the first institute. The subject matter was limited. There was a number (here four) of invited series of lectures by participants. The formal program ran to only about ten hours per week. Younger workers were welcome.

The budget of the first institute was $21,500, of which $20,000 came from the National Science Foundation and the remainder from the Society. There were twenty mathematicians present by invitation. The institute was open to all interested parties and nine others attended.

Almost immediately the Committee on Summer Institutes recommended that the institute be shortened to four weeks and that there be two topics, handled as two separate institutes simultaneously in the same location. The Council declined initially on both recommendations. At the fourth institute in 1956 it became clear that six weeks was indeed too long, as the program and the attendance became thin after four weeks. Thereafter, institutes were shortened to four weeks and then three.

Questions arose of the orientation of summer institutes and of who should be invited or admitted. The error in policy on admission and invitation had apparently been that the policy was at the same time too restrictive, in that the institutes were not properly publicized, and too lax, in that some people with the wrong expectations attended. The Council of 25 January 1961 made the following policy statement. "Summer Institutes should be research conferences, not primarily instructional in nature. But researchers at all levels of development should be admitted."

The growth and popularity of the institutes is exemplified in the twenty-second in 1974, where there were two hundred and seventy mathematicians in attendance, including eighty-four from foreign countries. The program there consisted of ten series of lectures and fifteen seminars. The budgeted cost of the institute was \$68,188, of which \$62,131 was covered by a grant from the National Science Foundation while the remaining \$6,057 was not recovered.

Because of the fact that the aim was research, the institutes themselves sometimes did not produce a published scientific record. Papers that grew out of work at the institutes were of course published in diverse locations. Lecture notes were prepared as a rule but by intent did not constitute publication. The Trustees of 29 November 1961 lodged the authority to produce and distribute them with the executive director with this instruction. "Complete distribution of Lecture Notes is to be made to all participants attending the institute or seminar, but beyond this only to individuals working in the particular field covered." However later Summer Institutes have yielded volumes in the series *Proceedings of Symposia in Pure Mathematics* (PSPUM, sometimes pronounced "PROVERBS" because it followed PSAM).

The committee in charge of an institute has been variously called an organizing committee and an invitations committee. It encompassed both kinds of duties and editorial duties as well. Members of the committee were usually among the principal speakers. Almost from the beginning each institute has had staff support from the Providence office consisting of a full-time employee doing preliminary work in the Providence office and work on site throughout the institute. The Providence office provided other support as did persons employed on the spot.

Here is a list of the Summer Institutes:

Lie Algebras and Lie Groups
    Colby College, 20 June–31 July, 1953
    Invitations committee: Nathan Jacobson, chairman, C. C. Chevalley, A. M. Gleason
    Publication: *Memoirs*, No. 14, consisting of five papers from the institute.

Several Complex Variables
  University of Colorado, 21 June–31 July 1954
  Invitations Committee: S. Bochner, chairman, D. C. Spencer, alternate
    chairman, L. Bers, N. Hawley, O. Zariski
  Editorial committee: S. S. Chern, W. T. Martin, Oscar Zariski
  Publication: *Bulletin* of The American Mathematical Society 62 (1956),
    79–141.

Set Theoretic Topology
  University of Wisconsin, 24 July–20 August 1955 [several versions of dates
    are in the files]
  Organizing committee: R H Bing, E. G. Begle, L. W. Cohen, G. T. Why-
    burn, R. L. Wilder

Differential Geometry in the Large
  University of Washington, 18 June–27 July 1956
  Organizing committee: C. B. Allendoerfer, chairman, H. Busemann, S. S.
    Chern, J. J. Stoker. (Sumner B. Myers had been named chairman but
    his unfortunate early death occurred in October 1955.)

Mathematical Logic
  Cornell University, July 1957
  Organizing Committee: P. R. Halmos, S. C. Kleene, W. V. Quine,
    J. B. Rosser, chairman, A. Tarski
  Publication: *Summaries of Talks Presented at the Summer Institute for
    Symbolic Logic*, published by the Institute for Defense Analyses.

Surface Area and Related Topics
  Bowdoin College, 16 June–11 July, 1958
  Organizing committee: T. Rado, chairman, L. C. Young, Lamberto Cesari,
    H. Federer, J. W. T. Youngs.

Number Theory
  University of Colorado, 21 June–18 July 1959
  Organizing Committee: Burton W. Jones, chairman, P. T. Bateman,
    A. Brauer, D. H. Lehmer, Ivan Niven, A. L. Whiteman.

Group Theory
  California Institute of Technology, 1–28 August 1960
  Organizing Committee: R. Brauer, R. H. Bruck, H. S. MacDonald Coxeter,
    R. P. Dilworth, M. Hall, Jr., chairman, H. J. Ryser
  Editor: M. Hall, Jr.
  Publication: PSPUM vol. 6. titled 1960 *Institute on Finite Groups.*

Applications of Linear Operator Theory
  Stanford University: 1–26 August 1961
  Organizing Committee: P. D. Lax, chairman, H. Helson, R. S. Phillips.

Relativity and Differential Geometry
University of California, Santa Barbara, 18 June–13 July 1962
Organizing Committee: A. H. Taub, chairman, Warren Ambrose, S. S. Chern, C. W. Misner

Differential and Algebraic Topology
University of Washington, 15 July–17 August 1963
Organizing Committee: N. Steenrod, chairman R. Bott, E. Dyer, J. Milnor.

Algebraic Geometry
Woods Hole, 6–31 July 1964
Organizing Committee: O. Zariski, chairman, W. L. Chow, M. Rosenlicht, D. C. Spencer, J. Tate.

Algebraic Groups and Discontinuous Subgroups
University of Colorado, 5 July–6 August 1965
Organizing Committee: A. Borel, co-chairman, G. D. Mostow, co-chairman, A. Selberg, T. Tamagawa
Editors: A. Borel and G. D. Mostow
Publication: PSPUM vol. 9.

Entire Functions and Related Parts of Analysis
University of California, San Diego, 27 June–22 July 1966
Organizing Committee: J. Korevaar, acting chairman, A. Beurling, R. Boas, chairman, Leon Ehrenpreis, W. Fuchs, L. Rubel.
Editors: S. S. Chern, Leon Ehrenpreis, J. Korevaar, W. H. J. Fuchs, and L. A. Rubel
Publication: PSPUM vol. 11.

Axiomatic Set Theory
University of California, Los Angeles, 10 July–5 August 1967
Organizing Committee: A. Robinson, chairman, P. Cohen, D. S. Scott
Editors: D. S. Scott and Thomas J. Jech
Publication: PSPUM vol. 13.

Global Analysis
University of California, Berkeley, 1–26 July 1968
Organizing Committee: S. S. Chern, cochairman, S. Smale, cochairman, F. Browder, L. Hörmander, I. Singer
Editors: S. S. Chern and S. Smale
Publication: PSPUM vol. 14–16.

Number Theory
SUNY at Stony Brook, 7 July–1 August 1969
Organizing Committee: D. Lewis, chairman, J. Ax, P. Bateman, K. Iwasawa, A. Selberg

Editor: D. J. Lewis
Publication: PSPUM vol. 20 under title 1969 *Number Theory Institute.*

Algebraic Topology
University of Wisconsin, 25 June–14 July 1970
Organizing Committee: A. Liulevicius, chairman, W. Browder, E. Fadell, E. Floyd, Peter Hilton, R. Lashof, M. Mahowald, R. Milgram, F. Peterson, J. Stasheff, E. Thomas
Editor: A. Liulevicius
Publication: PSPUM vol. 22.

Partial Differential Equations
University of California at Berkeley, 9–28 August 1971
Organizing Committee: Louis Nirenberg, chairman, Alberto P. Calderón, Lars V. Hörmander, Charles B. Morrey, Jr., James B. Serrin, Isadore M. Singer, Donald C. Spencer
Editor: D. C. Spencer
Publication: PSPUM vol. 23.

Harmonic Analysis on Homogeneous Spaces
William College, 31 July–18 August 1972
Organizing Committee: C. C. Moore, chairman, H. Furstenberg, Sigurdur Helgason, G. Hunt, B. Kostant, Robert P. Langlands, G. Mackey, E. Stein
Editor: C. C. Moore
Publication: PSPUM vol. 26.

Differential Geometry
Stanford University, 30 July–17 August 1973
Organizing Committee: S. S. Chern, cochairman, R. Osserman, cochairman, R. Bott, Eugenio Calabi, L. Green, Shoshichi Kobayashi, T. Milnor, B. O'Neill, J. Simons, I. Singer
Editors: S. S. Chern and Robert Osserman
Publication: PSPUM vol. 27.

Algebraic Geometry
Humboldt State University, 29 July–16 August 1974
Organizing Committee: D. Mumford, chairman, M. Artin, P. Griffiths, Robia Hartshorne, H. Hironaka, N. Katz
Editor: R. Hartshorne
Publication: PSPUM vol. 29 under title *Algebraic Geometry*—Arcata 1974.

Functions of Several Complex Variables
Williams College, 28 July–15 August 1975

Organizing Committee: Robert C. Gunning, cochairman, H. Rossi, cochairman, I. Craw, H. Grauert, D. Lieberman, J. Morrow, Raghavan Narasimhan, Yum-Tong Siu, R. O. Wells, Jr.
Editor: R. O. Wells, Jr.
Publication: PSPUM vol. 30.

Algebraic and Geometric Topology
Stanford University, 2–21 August 1976
Organizing Committee: R. Milgram, chairman, W. Browder, cochairman, E. Thomas, cochairman, R. Bott, P. Conner, R. Lashof, Robion C. Kirby, D. Quillen, William P. Thurston
Editor: R. James Milgram
Publication: PSPUM vol. 32.

Automorphic Forms, Representations, and L-Functions
Oregon State University, 11 July–5 August 1977
Organizing Committee: A. Borel, cochairman, W. Casselman, cochairman, P. Deligne, H. Jacquet, R. Langlands, J. Tate
Editors: A. Borel and W. Casselman
Publication: PSPUM vol. 33.

Harmonic Analysis in Euclidean Spaces and Related Topics
Williams College, 10–28 July 1978
Organizing Committee: S. Wainger, cochairman, G. Weiss, cochairman, D. Burkholder, A. Calderón, Y. Meyer, E. Stein, A. Zygmund
Editors: Guido L. Weiss and Stephen Wainger
Publication: PSPUM vol. 35 under title *Harmonic Analysis in Euclidean Spaces.*

Finite Group Theory
University of California at Santa Cruz, 25 June–20 July 1979
Organizing Committee: D. Gorenstein, chairman, J. Alperin, M. Aschbacher, N. Burgoyne, B. Cooperstein, G. Mason
Editors: Bruce Cooperstein and Geoffrey Mason
Publication: PSPUM vol. 37 under title *The Santa Cruz Conference on Finite Groups.*

Operator Algebras and Applications
Queen's University, 14 July–2 August 1980
Organizing Committee: R. Kadison, chairman, R. Douglas, E. Effros, R. Powers, L. Pukanszky, E. Woods
Editor: Richard V. Kadison
Publication: PSPUM vol. 38.

Singularities
Humboldt State University, 20 July–7 August 1981

Organizing Committee: P. Orlik, chairman, P. Church, A. Durfee,
  Martin Golubitsky, Le Duc Trang, P. Wagreich
Editor: Peter Orlik
Publication: PSPUM vol. 40.

Recursion Theory
  Cornell University, 28 June–17 July 1982
  Organizing Committee: A. Nerode, chairman, R. Shore, cochairman,
    S. Feferman, Yiannis N. Moschovakis, H. Putnam, G. Sacks, J. Schoen-
    field, R. Soare
  Editors: Anil Nerode and Richard A. Shore
  Publication: PSPUM vol. 42.

Nonlinear functional analysis and its applications
  University of California at Berkeley, 11–29 July 1983
  Organizing Committee: F. Browder, chairman, H. Brezis, Tosio Kato,
    J. Lions, L. Nirenberg, P. Rabinowitz
  Editor: Felix E. Browder
  Publication: PSPUM vol. 45.

Geometric Measure Theory
  Humboldt State University, 16 July–3 August 1984
  Organizing Committee: W. Allard, cochairman, F. Almgren, cochairman,
    E. Bombieri, R. Hardt, H. Lawson, Jr., J. Pitts, R. Shoen, W. Ziemer
  Editors: William K. Allard and Frederick J. Almgren, Jr.
  Publication: PSPUM vol. 44.

Algebraic Geometry
  Bowdoin College, 8–26 July 1985
  Organizing Committee: D. Eisenbud, chairman, S. Bloch, W. Fulton,
    D. Gieseker, J. Harris, R. Hartshorne, S. Mori
  Editor: Spencer J. Bloch
  Publication: PSPUM vol. 46.

Representations of Finite Groups and Related Topics
  Humboldt State University, 7–25 July 1986
  Organizing Committee: Jonathan Alperin, chairman, Charles W. Curtis,
    W. Feit, P. Fong
  Editor: Paul Fong
  Publication: PSPUM vol. 47.

Theta Functions
  Bowdoin College, 6–24 July 1987
  Organizing Committee: L. Ehrenpreis, cochairman, R. Gunning, cochair-
    man, E. Arbarello, D. Chudnovsky, G. Chudnovsky, T. Kawai,
    H. McKean
  Editors: L. Ehrenpreis and R. Gunning

Publication: PSPUM vol. 49.

Operator Theory/Operator Algebras and Applications
University of New Hampshire, 3–23 July 1988
Organizing Committee: William B. Arveson, cochairman, Ronald G. Douglas, cochairman, Ciprian I. Foiaş, I. C. Gohberg, Peter D. Lax, Donald Sarason, Barry Simon, Dan-Virgil Voiculescu.

SUMMER SEMINARS

The prototype of the Summer Seminar in Applied Mathematics was held at Brown University in 1941. In 1955 a similar event aimed at the immediate postdoctoral level was planned and the Society was asked to co-sponsor, at no expense, and to name a program committee. The Council agreed to the proposal including the naming of part of the program committee. In fact it developed that there was insufficient time to develop such a seminar for 1956. However the Council of 20 April 1956 voted to hold a Seminar on Mechanics in the Summer of 1957, consisting as initially proposed of two sessions of three weeks each. The level of participants was to be the Ph.D. The major objective was to be instructional. Nine lecturers were to give courses of about six lectures for three weeks while the second three weeks were to be devoted to individual lectures.

The Seminar as presented was shortened to four weeks. The topics were Fluid Mechanics, Solid Mechanics, Probability and Related Topics in the Physical Sciences, and Partial Differential Equations Including Numerical Methods. Support was obtained from the Air Force Office of Scientific Research, the Atomic Energy Commission, the National Science Foundation, the Office of Naval Research, and the Office of Ordnance Research, U. S. Army. The location was the University of Colorado.

Summer Seminars were held every two or three years until 1979 when they became an annual occurrence. Portions of many of the Summer Seminars in Applied Mathematics were published in the series *Lectures in Applied Mathematics* (LAM). Here is the list of Seminars:

Seminar in Applied Mathematics
Boulder, University of Colorado, 24 June–19 July 1957
Program Committee: M. H. Martin, chairman, Paul R. Garabedian, A. S. Householder, Mark Kac, R. E. Langer, C. C. Lin, William Prager, J. J. Stoker
Publications: LAM vols. 1A, 1B, 1C.

Modern Physical Theories and Associated Mathematical Developments
Boulder, University of Colorado, 24 July–19 August 1960
Organizing Committee: K. O. Friedrichs, chairman, M. Kac, M. M. Schiffer, G. E. Uhlenbeck, E. P. Wigner

Publications: LAM vols. 1, 2, 3, 4.

Space Mathematics
  Ithaca, Cornell University, 1 July–9 August 1963
  Organizing Committee: J. B. Rosser, chairman, D. Brouwer, M. S. Davis,
    W. R. Sears, V. T. Szebehely
  Editor: J. B. Rosser
  Publications: LAM vols. 5, 6, 7.

Relativity Theory and Astrophysics
  Ithaca, Cornell University, 25 July–20 August 1965
  Organizing Committee: A. H. Taub, chairman, S. Chandrasekhar,
    C. C. Lin, A. Schild, C. Misner
  Editor: J. Ehlers
  Publications: LAM vols. 8, 9.

Mathematics of the Decision Sciences
  Stanford, Stanford University, 10 July–11 August 1967
  Organizing Committee: G. B. Danzig, chairman, K. Arrow, R. H. Bush,
    R. Gomory, H. Kuhn, R. D. Luce, Robert Thrall, P. Wolfe
  Editors: G. B. Danzig and A. F. Veinott, Jr.
  Publications: LAM, vols. 11, 12.

Mathematical Problems in the Geophysical Sciences
  Troy, Rensselaer Polytechnic Institute, 6–31 July 1970
  Organizing Committee: W. H. Reid, chairman, H. Cohen, Richard C.
    DiPrima, D. Fultz, C. C. Lin
  Editor: W. H. Reid
  Publications: LAM vols. 13, 14.

Nonlinear Wave Motion
  Potsdam, Clarkson College of Technology, 5–28 July 1972
  Organizing Committee: Alan C. Newell, cochairman, C. Wilcox, cochair-
    man, N. Bleistein, V. Barcilon, D. Hector
  Editor: A. C. Newell
  Publication: LAM vol. 15.

Inverse Problems
  Los Angeles, University of California, Los Angeles, 5–16 August 1974
  Organizing Committee: V. Barcilon, cochairman, J. D. Cole, cochair-
    man, M. Crandall, F. Gilbert, L. Knopoff, R. G. Newton, J. Ralston,
    N. Grossman
  No publication.

Modern Modeling of Continuum Phenomena
  Troy, Rensselaer Polytechnic Institute, 7–18 July 1975
  Organizing Committee: R. C. DiPrima, chairman, George F. Carrier,
    H. G. Cohen, S. H. Davis, J. B. Keller, L. A. Segal

Editor: R. C. DiPrima
Publication: LAM vol. 16.

Nonlinear Oscillations in Biology
Salt Lake City, University of Utah, 12–23 June 1978
Organizing Committee: Frank C. Hoppensteadt, chairman, W. S. Childress, D. S. Cohen, P. Waltman, S. Winfree
Editor: Frank C. Hoppensteadt
Publication: LAM vol. 17.

Algebraic and Geometric Methods in Linear Systems Theory
Cambridge, Harvard University, 19–30 June 1979
Organizing Committee: R. W. Brockett, C. I. Brynes, C. Martin, S. K. Mitter, H. H. Rosenbrock, J. C. Willems
Editors: Christopher I. Byrnes, Clyde F. Martin
Publication: LAM vol. 18.

Mathematical Aspects of Physiology
Salt Lake City, University of Utah, 15–27 June 1980
Organizing Committee: F. Hoppensteadt, chairman, J. Rinzel, P. W. Altman, S. Stephenson, J. B. Keller
Editor: Frank C. Hoppensteadt
Publication: LAM vol. 19.

Fluid Dynamical Problems in Astrophysics and Geophysics
Chicago, University of Chicago, 29 June–10 July 1981
Organizing Committee: N. Lebovitz, chairman, V. Barcilon, R. DiPrima, F. Goldreich, J. Pedlosky, A. Toomre
Editor: Norman R. Lebovitz
Publication: LAM vol. 20.

Applications of Group Theory in Physics and Mathematical Physics
Chicago, University of Chicago, 6–16 July 1982
Organizing Committee: P. Sally, chairman, M. Flato, C. Fronsdal, I. Kaplansky, Y. Nambu, I. Singer, J. Wolf, G. Zuckerman
Editors: Moshe Flato, Paul Sally, Greg Zuckerman
Publication: LAM vol. 21.

Large-scale Computations in Fluid Dynamics
LaJolla, Scrips Institution of Oceanography, 27 June–8 July 1983
Organizing Committee: R. Somerville, chairman, A. Chorin, B. Engquist, S. Osher
Editors: Bjorn E. Engquist, Stanley Osher, Richard C. J. Somerville
Publication: LAM vol. 22.

Nonlinear Systems of Partial Differential Equations
Sante Fe, College of Sante Fe, 8–21 July 1984

Organizing Committee: B. Nicolaenko, chairman, D. Holm, J. M. Hyman
Editor: Basil Nicolaenko
Publication: LAM vol. 23.

Reacting Flows: Combustion and Chemical Reactors
Ithaca, Cornell University, 30 June–13 July 1985
Organizing Committee: G. S. S. Ludford, chairman, D. S. Cohen,
A. J. Majda, F. A. Williams
Editor: Geoffrey Ludford
Publication: LAM vol. 24.

Computational Solution of Nonlinear Systems Equations
Fort Collins, Colorado State University, 18–29 July 1988
Organizing Committee: E. L. Allgower, H. B. Keller, H. O. Peitgen,
W. C. Rheinboldt, S. Smale.

Symposia in Pure Mathematics

The Society had held symposia in pure mathematics sporadically almost from the beginning of its existence. These took the form of invitations to two or three people to lecture on closely related topics at the same meeting. A more formal arrangement of Symposia in Pure Mathematics was modeled on the Symposia in Applied Mathematics and began in 1959 with two symposia, one supported by the Institute for Defense Analyses. Proceedings of these symposia were frequently published in the series *Proceedings of Symposia in Pure Mathematics* (PSPUM).

Here is the list of symposia:

Lattice Theory
U.S. Naval Postgraduate School, 16–17 April 1960
Program Committee: R. P. Dilworth, chairman, G. Birkhoff, Alfred Tarski,
R. S. Pierce
Editor: R. P. Dilworth
Publication: PSPUM vol. 2.

Finite Groups
New York, 23–24 April 1959
Program Committee: A. A. Albert, chairman, Walter Feit, Marshall Hall,
Jr., Israel N. Herstein, Irving Kaplansky
Editors: A. A. Albert, I. Kaplansky
Publication: PSPUM vol. 1.

Differential Geometry
University of Arizona, 18–19 February 1960
Program Committee: C. B. Allendoerfer, chairman, Herbert Busemann,
Hans Samelson, D. C. Spencer

Editor: C. B. Allendoerfer
Publication: PSPUM vol. 3.

Partial Differential Equations
University of California, Berkeley, 21–22 April 1960
Program Committee: C. B. Morrey, Jr., chairman, David Gilbarg, Louis Nirenberg, Paul C. Rosenbloom
Editor: C. B. Morrey, Jr.
Publication: PSPUM vol. 4.

Recursive Function Theory
New York, 6–7 April 1961, jointly with the Association for Symbolic Logic and the Association for Computing Machinery
Organizing Committee: S. C. Kleene, J.C.E. Dekker, John McCarthy, J. B. Rosser, J. R. Schoenfeld
Editor: J. C. E. Dekker
Publication: PSPUM vol. 5.

Convexity
University of Washington, 13–15 June 1961
Organizing Committee: Victor Klee, chairman, David Gale, Branko Grünbaum, Merle Andrew, AFOSR liaison
Editor: V. Klee
Publication: PSPUM vol. 7.

Recent Developments in the Theory of Numbers
California Institute of Technology, 21–22 November 1963
Program Committee: A. L. Whiteman, chairman, Leonard Carlitz, D. H. Lehmer, W. J. LeVeque
Editor: A. L. Whiteman
Publication: PSPUM vol. 8 titled *Theory of Numbers.*

Singular Integrals
University of Chicago, 20–22 April 1966
Invitations Committee: Alberto Calderón, chairman, K. O. Friederichs, Robert T. Seeley, Antoni Zygmund
Editor: A. P. Calderón
Publication: PSPUM vol. 10.

Combinatorics
University of California, Los Angeles, 21–22 March 1968
Organizing Committee: T. S. Motzkin, chairman, Marshall Hall, Jr., Gian-
Carlo Rota
Editor: T. S. Motzkin
Publication: PSPUM vol. 19.

Applications of Categorical Algebra
New York, 10–11 April 1968
Organizing Committee: Hyman Bass, Alex Heller, chairman, John Moore
Editor: A. Heller
Publication: PSPUM vol. 17.

Nonlinear Functional Analysis
Chicago, 16–19 April 1968
Organizing Committee: Felix E. Browder, chairman, James Eells, Jr.,
Tosio Kato, George J. Minty, Richard S. Palais, Jacob T. Schwartz,
Stephen Smale
Editor: F. Browder
Publication: PSPUM vol. 18.

Representation Theory of Finite Groups
University of Wisconsin, 14–16 April 1968
Organizing Committee: Irving Reiner, chairman, Richard Brauer, Charles
W. Curtis, Walter Feit, James A. Green
Editor: I. Reiner
Publication: PSPUM vol. 21 under title *Representation theory of finite
groups and related topics.*

Analytic Number Theory
St. Louis University, 27–30 March 1972
Organizing Committee: Harold G. Diamond, chairman, Patrick S. Gallagher, Hugh L. Montgomery, Wolfgang M. Schmidt, Harold M. Stark
Editor: H. G. Diamond
Publication: PSPUM vol. 24.

Mathematical Developments Arising from the Hilbert Problems
Northern Illinois University, 13–17 May 1974
Organizing Committee: Felix E. Browder, chairman, Paul T. Bateman,
R. Creighton Buck, Donald J. Lewis, Daniel Zelinsky
Editor: F. E. Browder
Publication: PSPUM vol. 28.

Probability
University of Illinois, 15–18 March 1976
Organizing Committee: Kai Lai Chung, Joseph L. Doob, chairman,
Richard M. Dudley, Ronald K. Getoor, Frank B. Knight, Frank L. Spitzer
Editor: J. L. Doob
Publication: PSPUM vol. 31.

Relations between Combinatorics and Other Parts of Mathematics
Ohio State University, 20–23 March 1978

Organizing Committee: D. K. Ray-Chaudhuri, chairman, Marshall Hall, Jr., Peter J. Hilton, Gian-Carlo Rota, W. T. Tutte, Richard M. Wilson
Editor: Dijen K. Ray-Chaudhuri
Publication: PSPUM vol. 34.

Geometry of the Laplace Operator
University of Hawaii, 27–30 March 1979
Organizing Committee: Dalvid Bleecker and Robert Osserman, cochairmen, Victor Guillemin, Henry P. McKean, Jr., Karen Uhlenbeck, Joel Weiner, Alan Weinstein
Editors: Robert Osserman and Alan Weinstein
Publication: PSPUM vol. 36.

Mathematical Heritage of Henri Poincaré
Indiana University, 7–10 April 1980
Organizing Committee: Felix Browder, chairman, William Browder, Phillip A. Griffiths, Jürgen K. Moser, Stephen Smale, R. O. Wells, Jr.
Editor: Felix E. Browder
Publication: PSPUM vol. 39.

Several Complex Variables
University of Wisconsin, 12–15 April 1982
Organizing Committee: Yum-Tong Siu, chairman, Robert C. Gunning, F. Reese Harvey, Raghavan Narasimhan, Walter Rudin, Wilhelm F. Stoll, Shing-Tung Yau
Editor: Yum-Tong Siu
Publication: PSPUM vol. 41 under title *Complex analysis of several variables.*

Pseudodifferential Operators and Fourier Integral Operators with Applications to Partial Differential Equations
University of Notre Dame, 2–5 April 1984
Organizing Committee: Charles Fefferman, Victor W. Guillemin, Nancy K. Stanton, Michael E. Taylor, François Trèves
Editor: François Trèves
Publication: PSPUM vol. 43 under title *Pseudodifferential operators and applications.*

Mathematical Heritage of Hermann Weyl
Duke University, 12–16 May 1987
Organizing Committee: Michael F. Atiyah, Lipman Bers, Felix E. Browder, S. S. Chern, George D. Mostow, R. O. Wells, Jr., chairman, C. N. Yang
Editor: R. O. Wells, Jr.
Publication: PSPUM vol. 48.

Symposia on Mathematical Questions in Biology

On 5–7 April 1961 there was a symposium on Mathematical Problems in the Biological Sciences as part of the Spring Meeting in New York. The Invitations and Steering Committee included S. M. Ulam, chairman, R. E. Bellman, secretary, Anthony Bartholomay, John Jacquez, Theodore Puck, and C. E. Shannon. This was an isolated event and resulted in no consolidated publication.

At the meeting of the American Association for the Advancement of Science in Washington in December 1966, the Society presented a Symposium on Some Mathematical Questions in Biology under the direction of Murray Gerstenhaber. This appeared to be well received and on a motion by Professor Gerstenhaber the Council of 23 January 1967 agreed that a Committee on Mathematics in the Life Sciences should be appointed to propose topics for symposia from time to time and to recommend the constitution of the organizing committee in each instance.

The symposia became an annual event at meetings of the AAAS for many years. Each gave rise to a publication in the series titled Mathematics in the Life Sciences. The one in Washington was titled Some Mathematical Problems in Biology. Those that followed bore the title Some Mathematical Questions in Biology with a roman numeral through number X. The next three were unnumbered. Those that followed carried subtitles.

Effective in 1974, the Committee on Mathematics in the Life Sciences became a joint Committee with the Society for Industrial and Applied Mathematics. The AAAS became a sponsor of the publication of the symposia in 1977.

By 1986 it appeared that the audience for these symposia at the meetings of AAAS was not large enough to justify program time in the eyes of the management of AAAS although the books continued to sell and there was a segment of the biological community interested in quantitative problems. In April 1987 the Board of Directors of the Society for Mathematical Biology (SMB) agreed unanimously to cosponsor the annual Symposium on Some Mathematical Questions in Biology, with the understanding that it be held in alternate years at meetings of the Federation of American Societies of Experimental Biology (FASEB) and the American Institute of Biological Sciences. The last of the Symposia held with AAAS was the one in February 1987 and the first of those under SMB sponsorship took place at the FASEB meeting in May 1988.

Here are the location and editors of the symposia:

Washington, December 1966. Editor: Murray Gerstenhaber

I. New York, December 1967. Editor: Murray Gerstenhaber

II. Boston, December 1969. Editor: J. D. Cowan

   III. Chicago, December 1970. Editor: J. D. Cowan

   IV. Philadelphia, December 1971. Editor: J. D. Cowan

    V. Mexico City, June 1973, Editor: J. D. Cowan

   VI. San Francisco, January 1974. Editor: Simon A. Levin

  VII. New York, January 1975. Editor: Simon A. Levin

 VIII. Boston, February 1976. Editor: Simon A. Levin

   IX. Denver, February 1977. Editor: Simon A. Levin

    X. Washington, February 1978. Editor: Simon A. Levin

Houston, January 1979. Editor: Simon A. Levin

San Francisco, January 1980. Editor George F. Oster

Toronto, January 1981. Editor: Stephen Childress

Washington, January 1982. Subtitle: *Neurobiology.* Editor: Robert M. Miura

Detroit, May 1983. Subtitle: *Muscle Physiology.* Editor: Robert M. Miura

New York, May 1984. Subtitle: *DNA Sequence Analysis.* Editor: Robert M. Miura

Los Angeles, May 1985. Subtitle: *Plant Biology.* Editors: Louis J. Cross and Robert M. Miura

Philadelphia, May 1986. Subtitle: *Circadian Rhythms.* Editor: Gail Carpenter

Chicago, February 1987. Subtitle: *Models in Population Biology.* Editor: Alan Hastings

Las Vegas, May 1988. Subtitle: *The Dynamics of Excitable Media.* Editor: Robert Boyd

SUMMER RESEARCH CONFERENCES

The Summer Research Conferences instituted in 1980 by the Society with the cooperation of the Society for Industrial and Applied Mathematics (SIAM) and the Institute of Mathematical Statistics (IMS) were a byproduct of a change in stance on financial support of mathematics by the National Science Foundation (NSF). The general issue will be described in brief outline as background.

The success of the Institute for Advanced Study (IAS) and the AMS Summer Institutes prompted proposals for bringing mathematicians together in an atmosphere conducive to research. A historical account in the *Notices* of August 1978, pp. 481–492, consisting of material compiled by J. A. Krumhansl, then assistant director of the National Science Foundation, and followed by

Letters to the Editor, pp. 489–494, by Saunders Mac Lane and Joseph J. Kohn, were helpful in compiling this section. As early as 1968–1970 the Conference Board of the Mathematical Sciences considered versions of a summer research institute but did not proceed because there was no evident source for financing it. In 1969 the idea of a more permanent institute, that is, a replica of the IAS, was considered by the Advisory Committee of the Division of Mathematical Sciences of the NSF. A pressing consideration was the nature of the job market, in which young mathematicians nurtured in research were being employed at institutions in which the research atmosphere was not the prevailing one and research contacts were limited.

Alternate modes of support (these were catch words) were widely discussed. The Committee on Science Policy, then consisting of William J. LeVeque, chairman, R H Bing, Garrett Birkhoff, Felix E. Browder, John W. Jewett, Anil Nerode, and Elias M. Stein, reported to the Council in April 1976. The committee considered as possibilities the IAS model, the peripatetic institute (with continuing existence but moving from time to time from one host location to another), a greatly enlarged post-doctoral fellowship program, and short lived single subject institutes at sponsoring universities. The committee recommendation was for institutes of limited duration (say two plus years) in a subject field of a host institution, to be awarded competitively. It was approved. However, higher authorities did not take it up, although the NSF was aware of the report and the recommendation.

The National Science Board of March 16–17, 1978 did in fact approve the establishment of a Mathematical Sciences Research Institute and the issuance of a Project Announcement, i.e., a call for proposals. This was the competition that resulted in the founding of the Mathematical Sciences Research Institute in Berkeley and the Institute for Mathematics and its Applications at the University of Minnesota.

There was resistance in a portion of the mathematical community to alternate modes of support. The prevailing mode was the research grant that supplied summer support to a single mathematician or a small group, possibly including some students. The fear was that any other mode of support would come not from additional funds but from funds taken from the pool for summer support. Saunders Mac Lane was then a member of the NSB and tried valiantly, but without success, to explain to the mathematical community that either of those two viewpoints represents too great a simplification of a complex budget procedure. The resistance was directed specifically against the establishment of a permanent research institute.

When the project announcement for the proposed institute was circulated, it contained "an explicit addition stating that proposals for an institute would be evaluated in direct competition with other modes of support of research in mathematics." Some thought that this statement was not sufficiently prominent but the word circulated.

The Council of 23 January 1979 set an order of preference on some possible alternate modes of support. On a motion by Elias M. Stein the Council agreed to "set up a Special Committee on Modes of Support of Research to collect, examine, formulate and circulate proposals from the mathematical community to foster progress in research in mathematics and to make recommendations for appropriate action."

The committee consisted of G. D. Mostow, chairman, Herbert B. Keller, Calvin C. Moore, Ralph S. Phillips, James D. Stasheff, Elias M. Stein, and Hans F. Weinberger. At the Council of 20 April 1979 it made several recommendations, of which the first was that the Society "should submit a proposal to NSF for administration of an Oberwolfach-type institute which would arrange ten one-week conferences in specialized areas for about 30 participants."

The recommendation was approved by the Council and the proposal was accepted by the NSF for six conferences in the summer of 1980. Since then the norm has been ten. Occasional conferences have been of two weeks duration. Numbers of participants have been rather larger than the 30 suggested by the Committee, frequently being 50 to 60. A major part of the funds has come from the NSF but other government agencies have given support as well.

By 1983, the Summer Research Conferences were a joint venture with IMS and SIAM, in that the committee selecting the topics was a joint committee, first effective for the conferences of 1984, and there was a rough quota of conferences in core and applied areas. This change was made partly at the request of the NSF for their convenience. The Society has continued to solicit and accept grants, to administer the conferences, and to publish such proceedings as are appropriate.

Inasmuch as they are research conferences, much of the material presented is in the process of development. Publication of a volume is not a necessary outcome. In many cases, the work presented at these conferences is refined and presented in individual journal papers. However some conferences have found it reasonable to produce a volume of papers in the series *Contemporary Mathematics* (CONM), which is a vehicle for research monographs and conference proceedings.

In the following list, the time and place of the conference is given along with the subject and the name of the organizer. If there were proceedings, the volume number is stated. The editor is the same person as the organizer unless an editor is named. Note that in 1987 the conferences were in two locations.

1982 SUMMER RESEARCH CONFERENCES, University of New Hampshire, June 6 to July 17, 1982

1. Probabilistic computational complexity, Albert R. Meyer, organizer.

2. Ergodic theory and applications, Roy L. Alder, organizer.
3. Nonlinear partial differential equations, Joel A. Smoller, organizer. CONM 17.
4. Values of L-series at special points, Harold M. Stark, organizer.
5. Four-manifold theory, Robion C. Kirby, organizer. CONM 35. Cameron Gordon and Robion C. Kirby, editors.
6. Quantum fields, probability and geometry, Arthur M. Jaffee, organizer.

1983 SUMMER RESEARCH CONFERENCES, University of Colorado, Boulder, June 5 to August 13, 1983

1. Combinatorics and algebra, Richard P. Stanley, organizer. CONM 34. Curtis Greene, editor.
2. Applications of algebraic K-theory to algebraic geometry and number theory, Keith Dennis, organizer. CONM 55, 2 vol. Spencer J. Bloch, R. Keith Dennis, Eric M. Friedlander, and Michael R. Stein, editors.
3. Axiomatic set theory, James E. Baumgartner, organizer. CONM 31. James E. Baumgartner, Donald A. Martin, and Saharon Shelah, editors.
4. Group actions on manifolds, Reinhard Schultz, organizer. CONM 36.
5. Ordered fields and real algebraic geometry, D. W. Dubois, organizer.
6. Microlocal analysis, Linda Preiss Rothschild, organizer. CONM 27. M. Salah Baouendi, Richard Beals, Linda Preiss Rothschild, editors.
7. Fluids and plasmas: geometry and dynamics, Jerrold E. Marsden, organizer. CONM 28.
8. Probability theory, partial differential equations and applications, Daniel Stroock, organizer.
9. Geometrical analysis of singularities, Jeff Cheeger, organizer.
10. Kleinian groups, Howard Masur, organizer.

1984 SUMMER RESEARCH CONFERENCES, Bowdoin College, June 10 to August 18, 1984

1. New multivariate methods in statistics, Peter Huber, organizer.
2. Random matrices and their applications, Joel Cohen, organizer. CONM 50. J. E. Cohen and H. Kesten, editors.
3. The mathematics of phase transitions, Richard Timothy Durrett, organizer.
4. Aspherical complexes, Kenneth Brown and F. T. Barrel, organizers.
5. Group actions on rings, M. Susan Montgomery, organizer. CONM 43.
6. Diophantine problems, including diophantine equations, diophantine approximation, and transcendency, D. J. Lewis and W. M. Schmidt, organizers.

7. The Selberg trace formula and related topics, Audrey Terras, organizer. CONM 53. Dennis A. Hejhal, Peter Sarnak, and Audrey Anne Terras, editors.

8. Linear algebra and its role in systems theory, Biswa Nath Datta, organizer. CONM 47.

9. Integral geometry, Robert L. Bryant, organizer. CONM 63. Robert L. Bryant, Victor Guilleman, Sigurdur Helgason, and R. O. Wells, Jr., editors.

10. Complex differential geometry and non-linear differential equations, Y.T. Siu, organizer. CONM 49.

1985 SUMMER RESEARCH CONFERENCES, Humboldt State University, Arcata, California, June 23 to August 31, 1985

1. Brown-Gitler spectra and applications, R. James Milgram, organizer.

2. Applications of Lie groups in differential geometry, Wolfgang Ziller, organizer.

3. Numerical simulations of fluid flow, Gregory Baker, organizer.

4. Multiparameter bifurcation theory, Martin Golubitsky and John Guckenheimer, organizers. CONM 56.

5. Harmonic analysis in R^n, Elias M. Stein, organizer.

6. Function estimates, Murray Rosenblatt, organizer. CONM 59. J. S. Marron, editor.

7. Applications of mathematical logic to finite combinatorics, Stephen Simpson, organizer. CONM 65.

8. Combinatorics and ordered sets, Ivan Rival, organizer. CONM 57.

9. Current trends in arithmetical algebraic geometry, Kenneth A. Ribet, organizer. CONM 67.

10. Computational number theory, Andrew M. Odlyzko, organizer.

1986 SUMMER RESEARCH CONFERENCES, University of California, Santa Cruz, June 22 to August 2, 1986

1. Mathematics in general relativity, James Isenberg, organizer. CONM 71.

2. Large scale data analysis via computer graphics, Andreas Buja and Werner Stuetzle, organizers.

3. Time reversal of Markov processes and potential theory, Joseph Glover, organizer.

4. Artin's braid group, Joan S. Birman, organizer.

5. Discrete and computational geometry, Jacob E. Goodman and Richard Pollack, organizers.

6. Representation theory of Lie groups, Wilfried Schmid, organizer.

1987 SUMMER RESEARCH CONFERENCES

University of Colorado, Boulder, Colorado, June 14 to July 25 1987

1. Categories in computer sciences and logic, John W. Gray, organizer.

2. Hamiltonian dynamical systems, Kenneth Meyer and Don Saari, organizers.

3. Graphs and algorithms, Joseph P. Buhler and Phyllis Chinn, organizers.

4. Geometery of group representations, William Goldman and Andy Roy Magid, organizers. CONM 74.

5. The connection between infinite dimensional and finite dimensional dynamical systems, Basil Nicolaenko, organizer.

Cornell University, Ithaca, July 19 to August 15, 1987

6. Geometry of random motion, Richard Durrett and Mark Pinsky, organizers. CONM 73.

7. Crystal growth and pattern formation in phase transitions, Stuart P. Hastings and Nicholas D. Kazarinoff, organizers.

8. Complex analytic dynamics, John H. Hubbard, organizer.

9. Statistical inference from stochastic processes, Harahari U. Prabhu, organizer.

1988 SUMMER RESEARCH CONFERENCES, Bowdoin College, 11 June–5 August 1988

1. The mathematics and physics of order and disorder, Charles Radin, organizer.

2. Spatial statistics and imaging, Stuart Geman and Antonio Possolo, organizers.

3. Mathematical developments arising from linear programming, Jeffrey C. Lagarias and Michael J. Todd, organizers.

4. Geometric problems in Fourier analysis, William Beckner and Duong-Hong Phong, organizers.

5. Computational number theory, Carl Pomerance, organizer.

6. Current progress in hyperbolic systems, Riemann problems and computations, Barbara Lee Keyfitz and Brent Lindquist, organizers.

7. Mathematical problems posted by anistropic materials, Jean E. Taylor, organizer.

8. Geometric and topological invariants of elliptic operators, Jeff Cheeger and Alain Connes, organizers.

9. Elliptic genera and elliptic cohomology, Peter W. Landweber, organizer.

10. Control theory and multibody systems, Jerrold E. Marsden and J. C. Simo, organizers.

THE FINANCING OF CONFERENCES

Grants and contracts are the major source of the funds for conferences. A diversity of government agencies have given support. The National Science Foundation has been the largest and most frequent supplier of funds but the research supporting agencies of the Army, Navy, and Air Force have contributed as have other government groups including the National Security Agency, the Institute for Defense Analyses, the Atomic Energy Commission, and the Department of Energy. Some funds have come from foundations as well, including the Sloan Foundation, the Vaughn Foundation, and the IBM Foundation.

Grant money sometimes does not quite pay the full costs of a conference. There are costs of proposal preparation, expenditures made prior to the effective date of a grant, and unrecoverable overhead or unallowed costs. It is sometimes possible to cover certain kinds of costs from registration fees but on other costs this route is effectively barred. On the other hand, conferences lead to publications and these can be priced so that a modest conference deficit is in effect recovered.

There is an occasional symposium for which finally no grant is obtained. On such a conference the cash outlay of a few thousand dollars has been all built into the cost of the book. This is entirely fair in that the readers of the book benefit from the conference because without it there would have been no book.

Grants have become less generous with the passage of time. Honoraria for principal speakers, as opposed to travel costs and subsistence, have vanished and some agencies, notably the NSF, no longer allow manuscript fees. The latter, if desired, are a natural cost of a publication.

Supporting Journals

The Society publishes two journals that support the activities of the Society, namely the *Notices* and the *Abstracts*. The latter is a guide to the scientific content of meetings and a record of research while the former carries not only programs of meetings but also a variety of general information of professional interest.

The *Notices*

From the beginning of the Society through 1953, preliminary announcements and programs of meetings were issued as separata. Early issues seem to have vanished. The first in the files of the secretary (soon to be turned over to the archives) is for the Eighth Summer Meeting and Third Colloquium at Cornell University in Ithaca, NY in August 1901. Beginning with the 172nd Regular Meeting in New York on 31 October 1914, the collection is bound and gives the appearance of being complete.

In 1954, the *Notices* began as a periodical. One purpose was to take advantage of the favorable postal rates allowed to periodical publications. The issues were numbered consecutively. Volume numbers were added retrospectively, so that Issue No. 29 of February 1958 is also labeled Volume 5, Number 1.

The editorial management was initially lodged with the editors of the *Bulletin*, who were W. T. Martin and G. B. Price. However, after the first year, beginning with number 8 in February 1955, the editor was the executive director, then J. H. Curtiss. When the executive director changed from Curtiss to Gordon Loftis Walker in 1959, the editorship of the *Notices* changed as well. Volume 6, Number 5 of October 1959 was the first issue edited by Walker.

In addition to preliminary announcements and programs of meetings, the *Notices* contained notes on appointments and news about publications in mathematics. Announcements of fellowships, grants, and conferences appeared in increasing numbers. Abstracts of contributed papers were moved from the *Bulletin* to the *Notices* effective in 1958.

Letters to the Editor were a frequent subject of disagreement in the development of the *Notices*. The minutes of the Council of 29 December 1953 contain the following entry:

> The Secretary reported a request from a member of the Society that a "Letters to the Editor" column be established which would make it possible for members of the Society to state objections or clarifications to statements made in the Society's publications. The Council did not approve of this suggestion.

Not until 1958 did Letters to the Editor become a feature of the *Notices*. The first was a letter from R. P. Boas and G. Piranian concerning multiple submissions of an unspecified manuscript, accompanied by some remarks on the desired physical quality of manuscripts. The letter was followed by this note:

> The purpose of this new department is to provide a forum for discussions of the programs of the Society, and a method for communicating information of interest to the membership. Questions concerning matters of scholarship, such as those relating to the location of primary references, will be welcomed.

> The Council has instructed the Editor of the *Notices* not to allow the new department to be used for quick publication of mathematical results, and not to accept criticism of *specific individual papers* or of *specific individual reviews in Mathematical Reviews* or elsewhere. Thus a letter protesting against what the writer feels to be an unfair review in *Mathematical Reviews* of his or someone else's results is not eligible. There will be no editing of any letter, except for correction of obvious minor misprints, without direct negotiations with the writer.

The last paragraph of the note corresponds to the fact that the editor was solely responsible for the acceptance of letters.

In 1961, the Council voted to eliminate Letters to the Editor. When this was reported to the Business Meeting of 30 August 1961 in Stillwater, the meeting voiced its displeasure by passing a motion to place the matter on the agenda of the Business Meeting of 24 January 1962 in Cincinnati for discussion. In the intervening time, the Council reversed its position, as was reported in Cincinnati. At the same time, possibly to secure more diversified opinion on the suitability of Letters to the Editor, the bylaws were changed to lodge the editorship of the *Notices* with the executive director and the secretary.

This change in editorship was not enough. The Council came to regard the editors as too restrictive and too defensive in their handling of Letters.

Moreover, rejected Letters to the Editor were frequently coming to the Council on appeal. In 1976 an interim advisory committee was established. It was understood that it would not be concerned with abstracts, which are in the province of associate secretaries, or announcements but would be concerned with possibly controversial matters. In January 1976, the Editorial Board of the *Notices* was formalized as a committee elected by the Council in the following manner:

> The Editorial Board of the *Notices* shall consist of the executive director ex officio without vote, the Secretary ex officio with vote, and six other voting members with four year terms selected by the Council in the following manner. Every two years, three members' terms expire. The Nominating Committee submits a proposed slate of three to replace them, making an effort to achieve a Board broadly representative of different viewpoints within the Society on matters involved in *Notices* policy. The Council may ratify the whole slate, or else shall elect the three incoming Board members by preferential ballot. The first occasion when this procedure is followed, the three continuing members shall be obtained from among the present (appointed) Board.

This arrangement was quite satisfactory to the former editors, as it relieved them of considerable onus. At the same time, over the years, it brought into being a substantial administrative framework of circulating Letters to the Editor (and occasional articles and advertisements) and voting on them. However, the turmoil over Letters abated. The arrangement persisted except that the executive director was soon replaced by a deputy executive director, inasmuch as the editorial duty was not a direct part of the administrative management of the Society. It is noted that a slate has never been rejected.

At the Council of 28 December 1954, the secretary reported that a Survey of Research Potential and Training in Mathematics was to begin in January 1955. The committee supervising the survey consisted of A. A. Albert, chairman, L. Bers, L. W. Cohen, William L. Duren, A. Gleason, G. A. Hedlund, Mina S. Rees, R. L. Wilder, and S. S. Wilks. The investigator for the committee was J. W. Green. The Survey arose in the Division of Mathematics of the National Academy of Sciences/National Research Council, where Albert was division chairman. The division report of 1954–1955 shows J. A. Clarkson also as a member of the committee. Financial support came from the National Science Foundation through a contract with the University of Chicago and the committee submitted its reports to both the NSF and to the National Research Council.

The Albert Committee was no formal part of the apparatus of the Society. However, its work formed a basis for subsequent Society activity. In August 1956, President R. L. Wilder appointed a Committee to Investigate the Present Economic Status of Teachers, initially consisting of W. Givens, chairman, G. N. Garrison, and H. M. Schaerf. It was on the instigation of Schaerf that the Council mandated the appointment of the committee. In late 1957, a questionaire was sent to the sixty-one departments of mathematics included in the Albert Survey requesting minimum, median, and maximum salaries by rank for 1956–1957 and 1957–1958. Forty-two usable replies were obtained. Inasmuch as the year 1956–1957 was included, there was overlap with this aspect of the Albert Survey. The result was presented in the *Notices* for December 1957. This was the first edition of the Salary Survey, which has been published annually since then in the *Notices*.

Following the first survey, the committee was enlarged to include A. A. Albert, R. Bellman, D. Blackwell, and J. W. Green. The subsequent history is developed elsewhere.

From the beginning, meetings of other organizations and conferences that could be of interest to the membership were reported. Such items became more numerous either in actuality or in the reporting until these notices were collected into a Special Meetings Information Center beginning in February 1971. This was soon organized into a calendar of forthcoming meetings covering on the order of a hundred meetings spread over the next two years.

Abstracts of papers presented at meetings or offered by title were printed in the *Bulletin* after the fact for many years. The system was changed so that abstracts appeared currently in the *Notices*, effective with volume 5 (1958). Among the advantages was the fact that registrants at a meeting could better plan their time in attending competing sessions of papers. This arrangement continued through 1979, at which point the journal *Abstracts of the American Mathematical Society* was started.

From its beginning, the *Bulletin* was the journal of record of the Society. It contained reports of Council and Business Meetings, elections, reports of the treasurer, bylaws, and other items of Society business affecting the membership. The Council of 26 January 1977, in the course of considering the redirection of the *Bulletin*, ordered that the *Notices* become the journal of record of the Society.

The Committee on Membership (a committee of the Trustees), in looking for ways to make membership more attractive to persons who should naturally belong, proposed two changes to the Council of 20 April 1979. The first was approved in modified form "to endorse the principle of trying to make the *Notices* into an attractive mathematical newsletter which would be pleasant to read, and to reorganize it both technically and editorially to that effect." The second was approved as a recommendation to the Trustees that the *Notices* be separated into two journals, the prose and the abstracts.

Abstracts were only the beginning of the appearance of substantive mathematics in the *Notices*. It had always been regarded by many as a "throwaway" journal, of little value after the occurrence of the programs listed in it. Already in 1972, at the behest of the Committee to Monitor Problems in Communication, a column called "Queries" was instituted, with Wendell H. Fleming as the first associate editor. He was followed in 1975 by Hans Samelson, who served through 1987, and who was joined by Stuart Antman in 1985. The column "welcomes questions from members regarding mathematical matters such as details of, or references to, vaguely remembered theorems, sources of exposition of folk theorems, or the state of knowledge concerning published conjectures." Responses were supplied by readers. An effort was made to keep the column from becoming either a problem column or a vehicle for publication. The "Queries" column was consistent with the throwaway character of the journal at the time the column was initiated. By 1987 the "Queries" column appeared to have outlived its usefulness. During the next few months it was terminated.

With the prevalance of personal computers, there were two developments in the *Notices*. There was a series of articles, some written and some edited by Richard S. Palais, in a column called *Mathematical Text Processing*, of which the first appeared in January 1986. A column called *Mathematical Software*, edited by Joseph P. Buhler, also began in January 1986. It is too early to say what the stable state of either of these ventures will be.

When the journal became one of record, it was realistic to put articles with mathematical content of more permanent value in it. In 1982, Ronald L. Graham was named associate editor for special articles and the first such article, by J. C. Lagarias and titled *The van der Waerden Conjecture: Two Soviet Solutions*, appeared in the February issue. The new direction was a source of some discontent among those readers whose concept was still that of a throwaway.

The first advertisement in the *Notices* appears to be on the back cover of the issue of August 1960. (The date is inferred. Although the second class mailing data specify months of publication, the issues at the time were neither numbered nor dated.) It was an ad for four elementary textbooks from Holt, Rinehart and Winston. There were no more than half a dozen ads in the entire 1960 volume.

The page size was increased with the 1961 volume and the style was greatly improved. At the same time the amount of advertising gradually increased.

The pale brown cover appeared in 1964. Volume and issue numbers and dates, which had again disappeared, were restored, this being volume 11 of which number 1, part 1 was issue number 72 of January 1964.

Classified ads first appeared in 1976. The major category quickly became that for vacant positions. Advertising for situations wanted never became a large category, though it was free to the unemployed. Of course the employment register and the journal *Employment Information in the Mathematical Sciences* covered this area.

Effective in 1988, the *Notices* was remodeled. The page size was increased to $8\frac{1}{2} \times 11$ inches. The number of issues was increased from eight per year (seven in 1985 and 1986) to ten, one per month except for issues of May–June and July–August. Moreover, most of the material of the former December issue, principally the listing of assistantships and fellowships for graduate students, became a separate publication appearing earlier in the year. The revised *Notices* contains no more material than its predecessor but appears in more timely fashion.

Although the *Notices* has an editorial board, the burden of putting the book together issue by issue falls on the staff of the Society in Providence, that is, on the managing editor, the assistant to the managing editor, and persons who watch over the departments in the journal.

The editorial board of the *Notices* since 1976 has consisted of the secretary, ex officio, as chairman with vote, the managing editor without vote, and the following elected members:

Ed Dubinsky	6/76–1/83
Joseph B. Keller	6/76–1/77
Robion C. Kirby	6/76–1/81
Yiannis N. Moschovakis	6/76–1/77
Barbara L. Osofsky	6/76–1/81
Scott W. Williams	6/76–1/79
Arthur P. Mattuck	2/77–1/81
George Piranian	2/77–1/81
Richard J. Griego	2/79–1/83
M. Susan Montgomery	2/79–1/83
Ralph P. Boas	2/81–
Mary Ellen Rudin	2/81–
Bertram Walsh	2/81–1/85
Paul F. Baum	2/83–1/87
Raymond L. Johnson	2/83–1/87
Daniel Zelinsky	2/83–1/87
Steven H. Weintraub	2/85–
Robert L. Blattner	2/87–
Lucy J. Garnett	2/87–
Nancy K. Stanton	2/87–

The managing editors, ex officio, have been:

Gordon L. Walker	–10/77
William J. LeVeque	11/77–1/78

Lincoln K. Durst   2/78–1/85
Jill P. Mesirov   3/85–10/85
James A. Voytuk   11/85–

Abstracts

The journal *Abstracts* first appeared in 1980. It contained material that had appeared in the *Notices*. As its title implies, it contains abstracts of contributed papers, which are required as part of the process of contributing a ten-minute paper or participating in a special session, and abstracts of invited addresses, which are requested and are usually supplied but are not required. Abstracts appear in advance of the oral presentation except for some irregularities. Abstracts of papers presented "by title", that is, without oral presentation and not associated with a meeting are published as well. Beginning in 1982 abstracts from the Summer Research Conferences have appeared after the fact.

The journal *Abstracts* was "free" in its first year. Then a subscription charge was instituted and was gradually increased. Nonetheless, the journal is still subsidized from general funds.

The Editorial Committee of the *Abstracts* consists of the secretary and the associate secretaries. It is the latter who have supervised the acceptance of the contributed papers that are abstracted. The work of assembling the journal is done by the staff in Providence.

Books and Book Series

The Society has increasingly become a publisher of books. During the first fifty years, the Colloquium Publications were the only book series. The *Proceedings of the International Mathematical Congress of 1893* had been published by the New York Mathematical Society but there had been no other publication of books. The role of conferences in supporting several series of books has already been described. The place of translation in the scheme of publication by the Society has a chapter of its own. However, the first venture in book publication by the Society beyond the Colloquium Publications was the series Mathematical Surveys.

Mathematical Surveys

On 16 November 1940 there was a conference between the publisher Williams and Wilkins, represented by Messrs. Robert S. Gill and Passano, and the Society, represented by R.G.D. Richardson, T. C. Fry, and J. L. Walsh. The subject was the possibility that Williams and Wilkins publish surveys of fields of research in mathematics of current interest for the Society with editorial supervision by the Society but at no cost to the Society and as a nonprofit enterprise for the publisher, just as that publisher already did in other fields for other organizations. A model was the Ergebnisse series of Julius Springer. When the report of the conference was presented to the Council of 31 December 1940, it was voted to pursue the matter further and a committee consisting of E. W. Chittenden, chairman, T. H. Hildebrandt, H. P. Robertson, J. D. Tamarkin, and J. L. Walsh was appointed to do so. Walsh had already been mentioned at the conference as the possible editor.

The committee surveyed a sample of 180 members thought knowledgeable about such matters and the response was substantially favorable. Thus the Council of 2 May 1941 followed the recommendation of the committee that a second committee be appointed to confer with the company over the terms of an agreement. It was explicit that service on the negotiating committee not be a bar to subsequent service on the editorial committee.

The committee to confer consisted of Walsh, chairman, Chittenden, and E. J. McShane. At the Council of 2 September 1941, the new secretary, J. R. Kline, was added to the Committee. The agreement that was reached called for two books without further commitment on either side. However it was agreed subsequently that the use of the title Mathematical Surveys would rest with the Society.

The Company wished to use the phrase "Published for the American Mathematical Society by the Williams and Wilkins Company" but the Council preferred such a phrase as "Published by the Williams and Wilkins Company under the editorial supervision of the American Mathematical Society."

The temporary Editorial Board for Mathematical Surveys consisted of Walsh, chairman, A. A. Albert, and F. D. Murnaghan. However McShane replaced Walsh as chairman early in 1942, the latter having been called to active duty with the Navy. McShane in turn resigned before the end of 1942, when he took employment at the Ballistic Research Laboratories at Aberdeen Proving Ground. However before leaving he was able to announce that the first two volumes had been selected, namely *The Problem of Moments* by J. A. Shohat and J. D. Tamarkin and *The Theory of Rings* by N. Jacobson.

By October 1942, T. H. Hildebrandt had been appointed to the editorial committee to replace McShane and Murnaghan had been appointed chairman.

The manuscript of Jacobson came to about 128 pages and was priced at $3.50 with prepublication price of $2.75. The Council instructed the chairman to convey to the publisher the opinion that the price was rather high.

On 27 December 1942 the Council voted to publish the series Mathematical Surveys on its own responsibility. The list price of the Jacobson monograph was set at about $2.00 and of the Shohat–Tamarkin volume at about $2.50. In 1943 the price of each was set at $2.25.

Funds to initiate the new series came from the income of special funds, particularly the E. H. Moore Fund. The Council had voted to call the series the E. H. Moore Series of Mathematical Surveys. The Trustees at their meeting of 28 December 1942 cited unspecified complications and the Council of 27 February 1943 did settle on the name Mathematical Surveys.

When it was reported to the Council at the meeting of 12 September 1943 that the two books had been published, the Council voted to continue the series and to appoint a temporary editorial committee to serve until the by-laws could be amended to give the members of a standing committee the same Council representation as other editorial committees. The temporary committee in fact was Murnaghan, Albert, and Hildebrandt and the editorial committee did become a committee with Council representation effective 1 January 1945.

Publication in the series Mathematical Surveys continued at a leisurely pace. By 1984 the name of the Editorial Committee had been expanded to Mathematical Surveys and Monographs in response to a liberalization of the perception of the kinds of books to be published. By 1987 twenty-four volumes had appeared.

The members of the Editorial Committee of Mathematical Surveys have been the following:

A. A. Albert	1945
J. D. Tamarkin	1945
Nelson Dunford	1945–1949
A. W. Tucker	1946–1951
J. L. Walsh	1946–1950
W. T. Martin	1950–1952
Leo Zippin	1951–1956
R. J. Walker	1952–1957
M. Shiffman	1953–1955
I. J. Schoenberg	1956–1961
S. Ulam	1957–1962
I. Kaplansky	1958–1960
E. Hewitt	1961–1966
P. E. Conner	1962–1967
M. Suzuki	1963–1971
B. Yood	1967–1972
E. H. Brown, Jr.	1968–1976
R. G. Bartle	1972–1977
P. R. Halmos	1973–1975
M. A. Rosenlicht	1976–1978
R. J. Milgram	1977–1982
J. C. Scanlon	1978–1983
D. W. Anderson	1979–1984
R. O. Wells, Jr.	1983–
Gian-Carlo Rota	1984–1986
M. Susan Montgomery	1985–
Irwin Kra	1986–
Victor W. Guillemin	1987–

Contemporary Mathematics

In 1979 the charge to the Editorial Committee of the Surveys was enlarged. The Council of 23 January 1979 referred to the Committee on the Publication Program (a committee of the Trustees) a proposal from Raymond Ayoub that the *Transactions* promote special issues honoring outstanding mathematicians or obituary issues with analyses of the contributions of the deceased. The thought was that these would have a market for individual sales.

Supported by this referral, the Committee on the Publication Program recommended that the Society institute a series of books (dubbed informally the "Fast and Cheap Series") of proceedings of conferences and of lecture notes. The formal title proposed was Contemporary Mathematics. The Executive Committee and Board of Trustees of 11–13 May 1979 received this recommendation with favor and on their subsequent recommendation the Council of 22 August 1979 adopted this proposal. Editorial control was lodged with the Editorial Committee of Mathematical Surveys.

In the interval 1980–1987 there were sixty-seven volumes in the series Contemporary Mathematics. More than half have been reports of conferences or of special sessions. The series has received reports of many Summer Research Conferences. Editorial questions have been handled by a subcommittee of associate editors, with a member of the parent committee as managing editor.

The Council of 5 January 1988, on request of Irwin Kra, then managing editor, made the Editorial Committee of Contemporary Mathematics an independent editorial committee without Council representation.

OTHER PUBLICATIONS

The Society has accommodated others at the same time as it served its own constituency. An example is the Selected Tables in Mathematical Statistics, edited by the Institute of Mathematical Statistics but published by the Society, in which ten sets of tables were published since 1970. Another is the CBMS Regional Conference Series. These are expository softcover books sponsored by the Conference Board and supported by the National Science Foundation. Sixty-eight of these have appeared since 1970.

The Society has published the Conference Proceedings of the Canadian Mathematical Society, with eight volumes since 1982.

The AMS is the distributor in the United States, Canada, and Mexico of *Astérisque*, a journal of papers, lecture notes, and conference proceedings of the Société Mathématique de France that is sold also as separata.

The publication of the Society with the largest sale until recently is the *Russian–English Dictionary*, detailed in the chapter on translations.

Another very popular publication of the Society is *Mathematics into Type* by Ellen Swanson, long-time director of Editorial Services in the Society office. It began as an in-house manual for proofreaders, copy editors, and technical editors. It was rewritten in 1970 and published in 1971 for general consumption "by publishers and authors as a guide in preparing mathematics copy for the printer." With the change from Monotype as the standard for typesetting to phototypesetting, there were enough changes in practice to justify a revised edition in 1979.

There is a presage of the system of computer composition called T_EX in the revised edition of *Mathematics into Type* but the degree to which it was embraced by the Society does not yet appear. T_EX makes author prepared camera copy realistic and reduces or eliminates the problems of communication between author and printer addressed in *Mathematics into Type*.

T_EX was devised by Donald E. Knuth. His exposition with the title *The T_EXbook* was jointly published by Addison-Wesley and the Society in 1984. A user's guide for T_EX in the version called AMS-T_EX and sponsored by the Society was written by M. D. Spivak and published by the Society in 1986 under the title *The Joy of T_EX*. It has displaced the *Russian-English Dictionary* as the best seller.

Translations

Translation as a proposal appeared at the Council of 30 December 1947 when the Secretary reported that the Committee on the Role of the Society in Mathematical Publication was considering a project, to be funded by the Office of Naval Research (ONR), of translating articles in Russian. At the Council of 28 February 1948, A. W. Tucker reported formally for the committee. It was noted that a decision had been reported that the Russians would publish exclusively in Russian, making translation more necessary. The Council authorized the preparation of a list of technical terms supplemented by a pamphlet on the alphabet, phonetics, rules of syntax, etc. The Council agreed "[a]s a temporary measure [to] appoint a committee to be headed by Dr. [Ralph] Boas to serve as a committee of selection for articles (in Russian or other languages to be chosen at the discretion of the committee) of sufficient importance and demand to warrant translation into English and publication thereof." The position of Boas as executive editor of *Mathematical Reviews* was cited in making the appointment. Distribution of translations was referred to the committee and plans for funding beyond an initial approach to ONR for $25,000 were left unresolved. At the Council of 7 September 1948 it was reported that the contract with ONR to produce both the translations and the pamphlet had been signed. Immediately following WWII, the ONR was carrying part of the burden that would eventually fall to the National Science Foundation when that body was established.

In December 1948 the Committee, consisting of R. Boas, chairman, S. Eilenberg, D. H. Lehmer, W. Prager, and G. Y. Rainich, reported that the translation of four papers had been authorized, of which one had been completed except for final editing. In April 1949, the first set was ready for distribution. The arrangement was that 200 copies were to go to the Navy for their distribution and that additional copies would be sold by the Society at cost. However the Trustees found the distribution costs for the Society to be too high. Forty-two papers were in various stages of translation by August 1949. The project was temporarily halted in 1950 with fifteen papers translated. The word list of 2800 words and the pamphlet on grammar of about 30 pages was compiled by a committee consisting of S. Eilenberg, chairman,

L. Bers, P. R. Halmos, G. Y. Rainich, and A. E. Ross and was ready for distribution in 1950. This was a forerunner of the *Russian-English Dictionary* (the Lohwater dictionary) discussed later.

In 1952 the original committee was discharged and a new committee with staggered terms was established, consisting of R. E. Bellman, R. P. Boas, chairman, J. L. Doob, I. Kaplansky, and H. Samelson.

In 1953 it appeared that the ONR, though satisfied with the translation project, no longer had the funds to continue to support it. The National Science Foundation (NSF) was asked to provide support, which it subsequently agreed to do.

The Council of 30 April 1954 considered whether to continue to distribute translations as separata or to consolidate them in volumes. Unable to reach an effective decision the Council referred the matter to the Executive Committee with power. The Executive Committee approved the change on 24 June 1954. One hundred five papers had appeared under the ONR sponsorship and were reprinted in hard cover volumes with NSF support in 1962. This is the book series AMS Translations-Series 1. The AMS Translations-Series 2 has put out 136 volumes between 1955 and 1987. Support from the NSF ended in 1966 but the venture by the Society has been self-supporting.

In December 1955 the Translations Committee reported that the reservoir of the older material in need of translation was almost exhausted so that for the most part only current material was being translated.

The report of the Committee on Translations for 1956 notes that along with the papers translated from Russian there was one from Japanese.

In 1959 the Committee on Translation from Russian and other Foreign Languages became a joint committee with the Institute of Mathematical Statistics with the addition of E. Lukacs and I. Olkin to the committee. The committee operated as two subcommittees. The Association for Symbolic Logic became a participant in 1983 with the appointment of a subcommittee consisting of S. Feferman, J. P. Jones, V. Lifschitz, and G. Minc.

In September 1959 the Council voted to expand the translation program to include the entire mathematics section of *Doklady Akademii Nauk SSSR*, beginning in 1960, under the title *Soviet Mathematics-Doklady*. Over the years, six others followed, namely:

> *Trudy Moskovskogo Matematicheskogo Obshchestva* under the title *Transactions of the Moscow Mathematical Society* since 1963. This is a joint arrangement with the London Mathematical Society.

> *Trudy ordena Lenina Matematicheskogo instituta imeni V. A. Steklova* under the title *Proceedings of the Steklov Institute of Mathematics* since 1965.

Izvestiya Akademii Nauk SSSR Seriya Matematicheskaya under the title *Mathematics of the USSR-Izvestiya* since 1967.

Matematicheskiĭ Sbornik under the title *Mathematics of the USSR-Sbornik* since 1967.

The mathematics part of *Vestnik Leningradskogo Universiteta* [Seriya] Matematika, Mekhanika, Astronomiya under the title *Vestnik Leningrad University*: Mathematics during 1974–1984.

Teoriya Veroyatnostei i Matematicheskaya Statistika under the title *Theory of Probability and Mathematical Statistics* since 1974.

Beginning with *Izvestiya*, these translations were done without contract support.

The translation of *Acta Matematica Sinica* under the title *Chinese Mathematics-Acta* was established by the Council of January 1961 but for political reasons lasted only through 1967 when the Cultural Revolution took hold. At the time that the translation began, S. H. Gould was near the end of his tenure as executive editor of *Mathematical Reviews*. Although he had taught and written in mathematics, his Ph.D. was in classics and the study of a diversity of languages was a consuming interest and lay in both the written and the spoken language. Following his tenure at *Mathematical Reviews* he became the first editor of Translations and served from 1960 to Summer 1972. He also served as editor of the translation journal. The project received financial support from the National Science Foundation.

Although rather few books have been translated as part of the program of translation from Russian, this is a circumstance of the business rather than a deliberate decision. Commercial publishers have acquisition editors who seek and contract for translation of books in the manuscript stage whereas the Society has waited for printed books to review them. Officers of the Society would like to have more translation of books done through the Society because of a conviction that the Society can serve the profession by producing higher quality translations at lower cost. Whether the Society can and should enter the market at the manuscript stage was an open question for many years but in 1987 the Executive Committee and Board of Trustees decided to try it beginning in 1988 through the good offices of distinguished Russian consultants to the Committee on Translation. The first of these were V. I. Arnol'd, S. G. Gindikin, and N. K. Nikol'skiĭ.

At the time of preparation of this volume the political situation in China had been relaxing for the past ten years and mathematics was once again flourishing. Arrangements were accordingly in place again to translate mathematical work from Chinese to English. The project is supervised by a committee consisting of Tsit-Yuen Lam, Chairman, Sun-Yung Alice Chang,

S. Y. Cheng, Tai-Ping Liu, and Chung-Chun Yang. The first two books are anticipated in 1988.

Arrangements for the translation of mathematics from Japanese into English were also being made. This project was supervised initially by an ad hoc committee consisting of Nagayoshi Iwahori, Tosio Kato, Shoshichi Kobayashi, Masayoshi Nagata, and Katsumi Nomizu. In the stable arrangement the committee consists of Nomizu and Kobayashi with consultants as needed. In 1988 a contract was signed with the Mathematical Society of Japan and Iwanami Shoten, Publishers, for the translation of the expository articles in the journal *Sūgaku* under the title *Sūgaku Expositions*.

Royalties to foreign authors of translated books were authorized in 1961 even though there may have been no copyright or obligation to pay. Over the years, there was a difficult issue, namely how to pay a Russian author individually rather than his government. The general principle adopted by the Trustees in 1962 was that funds were held until an occasion when the author or a personally accredited representative was traveling in the United States. As a result, there is a fund of unpaid royalties awaiting claimants. This situation changed when the USSR adhered to the international copyright agreement since the right to receive payment lodges with the copyright holder.

Although the start-up costs of a translation journal are substantial, with an investment to be made before a subscription list is established, in the long run it is possible to operate such a journal at a surplus without charging excessively large prices.

RUSSIAN-ENGLISH DICTIONARY

The Society has published a few items whose sales far exceed the rest. Among them is the *Russian-English Dictionary of the Mathematical Sciences*, which by 1987 had sold over 9600 copies. There is a parallel English-Russian dictionary published by the Foreign Languages Publishing House in Moscow.

The foreword and the preface to the Soviet edition contain a brief history of the project. Informally, it arose in June and July 1956 when N. E. Steenrod and A. J. Lohwater were the American delegation to the Third All-Union Congress of Soviet Mathematicians in Moscow. Formally, there was an exchange of letters between D. Bronk, president of the National Academy of Sciences, and A. N. Nesmeyanov, president of the Soviet Academy of Sciences, followed by the appointment of representative committees. The Soviet committee consisted of P. S. Alexandrov, chairman, I. R. Shafarevich, M. M. Postnikov, A. F. Leont'ev, and A. A. Dezin, later augmented by L. D. Kudryavtsev, vice-chairman, L. N. Bolshev, S. M. Nikol'skii, E. D. Solomentsev, and V. S. Vladimirov. The American committee consisted of S. H. Gould, A. J. Lohwater, and G. Y. Rainich.

Work began with a meeting of representatives of the two committees on 10 September 1958 at the Steklov Institute in Moscow. Coverage was intended for use in reading papers and books at any level in mathematics and selected areas of theoretical physics, defined by the coverage of *Mathematical Reviews* and *Referativnyĭ Zhurnal Matematika*. Each dictionary contains a brief treatment of the appropriate grammar. Each side prepared a word list and the two lists were exchanged for critical comment.

The Russian-English dictionary was prepared by Lohwater with the collaboration of Gould, who was then executive editor of *Mathematical Reviews*. Financial support came from the National Science Foundation. Work at this end was completed by January 1960 and the Soviet suggestions for improvement were incorporated by November 1960. The book of xiii+267 pages was published in 1961. It contains more than 15,000 terms.

The first edition of such a book is naturally not quite perfect. As a result, frequent users have compiled private word lists to supplement it. The advances in the field of mathematics have brought new vocabulary into use. The 1961 dictionary was weak in terms from applied mathematics and statistics and has little of the vocabulary of computer science, whose development was only beginning in 1960. Thus in 1984 exploration of a revision was begun. An advisory committee consisting of H. Weinberger, chairman, R. C. Bartle, P. R. Halmos, and E. Hewitt was formed, with R. Boas as consultant. The National Security Agency (NSA) was approached for funds and expressed considerable interest but with anticipated budget cuts did not have the money then.

In a trip to Russia in 1986 W. J. LeVeque, the Society's executive director, talked with people at VINITI (All-Union Institute of Scientific and Technical Information) and at the Steklov Institute about a revision and found a positive response. The Trustees agreed to embark on the venture without assurance of full funding but with a limitation on expenditure and scope of the project should funding not be forthcoming. The NSA was approached in April 1987 for a two-year project of preparation of the dictionary, with R. P. Boas as editor, and did provide an initial grant of $200,000. Thus the project is under way.

Publication Problems

There has been a sequence of committees concerned with publication. When the only publications of the Society were the *Bulletin*, the *Transactions*, and the Colloquium Publications, editors were themselves closely concerned with printing contracts and other questions of publication along with the Trustees. There was a Committee on Printing Contracts that was reconstituted in 1942 and then consisted of Tomlinson Fort, chairman, Mark H. Ingraham, and J. D. Tamarkin. It acted on consultation with editorial committees on all printing and made recommendations to the Trustees directly. R. M. Foster replaced Fort in 1944 and became chairman.

At its meeting of 23 November 1945 the Council approved a recommendation of the Committee:

> That the Office Manager of the Society be authorized to conduct in cooperation with the committee on Printing Contracts a thorough investigation of the possibilities of various alternative methods to conventional printing from type,—in particular, the photo-offset reproduction of specially prepared type-written copy.

The Colloquium Publication *Lectures on Matrices* of J. M. Wedderburn had already been reprinted in 1944 by photo-offset by Edwards Brothers. It was the use of type-written copy that required investigation.

In December 1946, the Committee on Printing Contracts reported that initial investigation showed that the cost of the *Bulletin* produced in the proposed manner would be slightly higher and that of the *Transactions* was at most negligibly lower.

In April 1949 a Committee to Study Special Methods of Reproducing Mathematical Research was appointed, consisting of B. P. Gill, chairman, S. Eilenberg, and P. A. Smith. The newly proposed *Memoirs* were the vehicle proposed for the experiment.

The executive director, in place since the end of 1949, was made an ex officio member of the Committee on Printing Contracts.

There were other ad hoc committees on the general problems of publication associated with the substantial increase after WWII of both the amount of material offered for publication and the cost of publishing it. One was a Committee to Study the Problems of Publication, consisting of A. W. Tucker, chairman, E. Hille, J. R. Kline, and P. A. Smith. At the Council of 27 December 1951, it recommended among other things "[t]hat various present committees on non-editorial matters be consolidated in a single Publications committee consisting of three members elected for three year terms in rotation." The Council approved and the preparation of a change in the bylaws was authorized.

COMMITTEE ON PRINTING AND PUBLISHING

The Committee on Printing and Publishing was formally approved in September 1952 and in December the various ad hoc committees and the Committee on Printing Contracts were discharged.

The Committee on Printing and Publishing were members of the Council. They were the following:

M. Hestenes	1/53–12/54
J. R. Kline	1/53–7/55
C. J. Rees	1/53–12/56
E. F. Beckenbach	1/55–12/60
A. W. Tucker	8/55–12/61
E. G. Begle	1/57–12/62
G. A. Hedlund	1/61–12/66
J. D. Swift	1/52–12/67
C. W. Curtis	1/62–12/68
G. Seligman	1/67–12/69
L. Auslander	1/68–12/70
F. E. Browder	1/69–12/71

Publication processes grew more complex. Operations, particularly composition, were increasingly done in-house as techniques and equipment were developed. The staff took on a larger share of the work and management of publication. For example, in 1955 the Council voted that printing contracts should be made by the executive director on the basis of policy laid down by the Committee on Printing and Publishing.

The committee did concern itself with subventions to outside journals and with the problem of assuming publication or editorial control of *Mathematics of Computation*. Both of these issues are detailed elsewhere. Problems connected with page charges went to the committee also.

COMMITTEE TO MONITOR PROBLEMS IN COMMUNICATION

An ad hoc Committee on Information Exchange and Publication in Mathematics, consisting of D. Zelinsky, chairman, P. R. Halmos, Albert Madansky, Alex Rosenberg, J. Barkley Rosser, J. D. Swift, J. F. Traub, and A. S. Wightman, and made its final report to the Council of 23 January 1967. It was concerned with a variety of matters such as preprint exchanges, publication by abstract with available filed manuscript, problems of information retrieval, communication with other disciplines, expository books, and conferences, some of which were to be objects of future attention.

One of its recommendations, adopted by the Council, was to establish the Committee to Monitor Problems in Communication with the charge

> that this Committee should experiment with pilot projects with (hopefully) improved methods of communication if such projects are approved by the Society and other organizations concerned, and that the results of these experiments be carefully assessed after a suitable trial period.

The committee was established in 1967. When its chairman was about to be given Council membership, the secretary observed considerable overlap between its charge and that of the Committee on Printing and Publishing. On his suggestion, the Council of 27 August 1968 agreed to phase out the latter committee.

The initial membership of the committee was W. J. LeVeque, chairman, F. Browder, I. Niven, A. Rosenberg, and D. Zelinsky. J. Douglas, Jr., and G. L. Walker were soon added.

Comm.-Comm., as it is universally known, has considered a diversity of products of conceivable benefit to the mathematical research community, a few of which are mentioned here.

The first recommendation of the committee was to recommend a Conference on Communication Problems to take place as the new headquarters was dedicated. It was here K. Chandrasekharan proposed the idea of a current awareness journal. The Committee to Monitor Problems in Communication picked up on the suggestion, which went through the formal route of approval. The journal appeared in 1969 under the title *Contents of Contemporary Mathematical Journals*; it consisted simply of photoreproductions of the tables of contents of current issues of many of the major journals in mathematics and was very inexpensive. The format was changed by sorting the articles from various journals by subject according to the AMS(MOS) subject classification, described below, when that was in place. The title was subsequently changed to *Current Mathematical Publications*. The journal had a separate editorial committee until 1977, when the task was assigned to the Editorial Committee of *Mathematical Reviews*. It is prepared from incoming information at *Mathematical Reviews* and is incorporated into the mechanized process in such fashion that bibliographic information need be entered only once for both journals.

One of the most innovative projects which the AMS undertook in the 1960s, on recommendation from Comm.-Comm. to the Council of 29 August 1967, was the Mathematical Offprint Service (MOS), developed under a grant from the NSF. MOS began operation in 1968 and was a subscriber-based information retrieval service for the mathematical community. MOS was one of the first attempts to provide selective dissemination of information in a scientific discipline. It was modeled after an internal system at Bell Telephone Laboratories and was very successful with the portion of the community which it served. This account of MOS was prepared by William B. Woolf.

MOS subscribers filled out detailed "profiles" of their interests to be matched against descriptions of articles from more than 200 worldwide journals. In most cases, the journals provided the AMS with advanced proofs of articles they were about to publish, adding a very current aspect to the information MOS was providing. Full bibliographic information about each article was added to a computer database of information to be compared regularly by computer with subscriber profiles. Based on the degree of match, subscribers might be sent actual copies of articles, or author-supplied abstracts, or bibliographic listings.

The cost to subscribers was unbelievably low! Initially an offprint cost $.30, regardless of its length, and a title listing cost $.03. Over time, the prices were increased somewhat but never enough to allow MOS to operate in a way that convinced the AMS and NSF that the project would become self-supporting.

When MOS lost NSF support in 1971, it was replaced by a multi-part system made up of a Mathematical Title Service (similar to MOS but with no offprints), *Index of Mathematical Papers* (a semi-annual publication containing an author index and a subject index to the papers processed by MTS) and an ever-growing data base of article information. This system was also NSF-supported.

During four years of operation, MOS (and MTS) popularity grew until it had a subscriber base of more than 1000 mathematicians. At the time it ceased operation, it was processing more than 800 articles a month.

One of the most long-lasting products of MOS was the AMS(MOS) Subject Classification Scheme (1970). When MOS began operation, subjects were designated by four-digit classification numbers taken from the current Subject Classification. However, a massive redesign of this subject classification was done during 1969/1970, resulting in the tri-level AMS(MOS) Subject Classification (1970). This was not only used by MOS and by AMS journals but was eventually adopted by *Mathematical Reviews* and was the basis for the classification which *MR* and AMS primary journals presently use.

In retrospect, MOS could be judged to have been ahead of its time. Over the years, the need that MOS fulfilled has been met by a host of other services tailored somewhat to meet individual needs. These include services such as MathSci, subscriptions to individual sections of *Mathemathical Reviews*, and sales of article reprints by *MR*. The *Index of Mathematical Papers* eventually was replaced by *MR*'s own several indexes, by subject and author, covering varying periods of time.

The use of classification numbers with papers published in Society journals is required and has spread widely to other journals and to books. In the last, it is an aid to librarians.

Repeated offers to assist with a revision of the Dewey Decimal Classification of mathematics, universally recognized as ineffective and outdated for research mathematics, have been refused.

Following a recommendation from Comm.-Comm., in several fields reviews from *Mathematical Reviews* have been collected and in some cases reclassified in light of subsequent development and have been ordered to make it possible to follow a developing theory easily. The prototype is the two volume set titled *Reviews of Papers in Algebraic and Differential Topology, Topological Groups, and Homological Algebra* [1940–1967] edited by Norman E. Steenrod. The success of such a project depends on finding an editor who is eager to undertake the task in a popular field. The Steenrod volume contained more than 6000 reviews. A much larger project of the same kind, by far the largest in this area, is the one in number theory, consisting of 14426 reviews in six volumes titled *Reviews in Number Theory* [1940–1972], edited by William J. LeVeque, followed by 17119 reviews in six volumes titled *Reviews in Number Theory* 1973–83, edited by Richard K. Guy. Other fields covered to date are K-theory, ring theory, graph theory, finite groups, infinite groups, numerical analysis, partial differential equations, and global analysis. The degree of editing with respect to reclassification and ordering varies among collections.

The committee recommended in 1980 the employment of a staff writer in the Providence office. This was achieved in 1986, with substantial benefit in informing the mathematical community.

The committee was responsible for the 1983 Survey of American Journal Prices, which was the first in a series.

As the publication program of the Society increased and diversified, Comm.-Comm. was charged in August 1985 to be the editorial committee for books not in series.

Not all of the ideas in this direction have come from Comm.-Comm. There has been a Committee on the Publication Program responsible nominally to the Trustees that has made a variety of suggestions about books and collections that are in the process of preparation as this history is written.

The successive chairmen of Comm.-Comm., all but the first of whom were Council members, are the following:

W. J. Leveque	2/67–12/70
A. Rosenberg	1/71–12/74
Leonard Gillman	1/75–12/76
D. W. Bailey	1/77–12/78
G. B. Seligman	1/79–12/79
P. T. Church	1/80–12/80
R. G. Bartle	1/81–12/82
W. W. Comfort	1/83–12/83
Lynn A. Steen	1/84–12/84
M. B. Pour-El	1/85–

The issue of alternative methods of composition, such as the use of varitype already mentioned in connection with the inception of the *Memoirs*, came up repeatedly. The Society experimented with a diversity of equipment, introducing it wherever it could be done without encountering the preference of the members for what looked like cold type. As equipment improved this was increasingly possible. The versatility of TeX has made its use possible for all purposes of composition and essentially all composition is now done in the Providence or Ann Arbor offices. The most elaborate such project was the computerization of the production of *Mathematical Reviews*, which is described in some detail in that chapter.

International Congresses

THE CONGRESS OF 1950

The AMS has been concerned with several international congresses. At the International Congress of 1936 in Zurich, the invitation from the United States to the Congress to convene in America in 1940 was accepted. It was planned to hold it in the interval 4–12 September 1940 in Cambridge, MA, at Harvard University and the Massachusetts Institute of Technology. The Society was to be the entrepreneur.

The following appointments were proposed by the Nominating Committee and approved by the Council:

President:	G. D. Birkhoff
Organizing Committee:	W. C. Graustein (chairman), G. D. Birkhoff, G. A. Bliss, G. C. Evans, J. R. Kline, Solomon Lefschetz, Øystein Ore, R.G.D. Richardson, M. H. Stone, J. L. Synge, J. D. Tamarkin, J. M. Thomas, Norbert Wiener
Editorial Committee:	Einar Hille (chairman), F. D. Murnaghan, P. A. Smith
Financial Committee:	Marston Morse (chairman), J. L. Coolidge, M. H. Ingraham, H. B. Phillips
Secretary:	R.G.D. Richardson.

Arrangements went forward through 1937 and on into 1939 in the expected manner. However, the attack of Germany on Poland on 1 September 1939, with the declaration of war on Germany by Britain and France, altered plans abruptly. At the Council of 6 September 1939, Professor Graustein, speaking for the Organizing Committee, proposed suspension of activity, postponement of the Congress to a more favorable time, and the appointment of an Emergency Executive Committee to announce postponement of the Congress at a time that it should determine and to take the initiative for resumption of activity. Moreover, invited speakers were to be notified immediately both of the possibility of postponement and of the termination of invitations to speak in the event of postponement. These proposals were adopted, as were proposals aimed at conserving the contributions and pledges of funds for the Congress when it was rescheduled.

The Emergency Executive Committee consisted initially of G. D. Birkhoff, W. C. Graustein, Einar Hille, M. H. Ingraham, J. R. Kline, Marston Morse, R.G.D. Richardson, and M. H. Stone. With the death of Birkhoff and of Graustein, T. H. Hildebrandt and D. V. Widder were appointed.

At the Council meeting of 26 April 1946, when World War II was over, the Emergency Executive Committee, of which Morse was then chairman, reported that "it was interested in the revival of plans for the Congress only if that Congress could be an open Congress to which all mathematicians would be invited, irrespective of national allegiance." The committee was of the opinion that 1948 was too soon but that 1950 should be explored as a date for holding an open Congress.

The motion that established the Emergency Executive Committee had stipulated that when its recommendation to proceed with a Congress was adopted it should be automatically discharged and that the original committees of the Congress should be reactivated. The Council, in approving the report of the committee, amended its previous action so that officers and committees of the Congress were to be appointed and the Emergency Executive Committee automatically discharged.

At the Council of 26 April 1947, the invitation from Harvard University to hold the Congress of 1950 there was accepted.

On 28 February 1948, a nominating committee under the chairmanship of Professor D. V. Widder recommended officers and committees for the Congress as follows:

Secretariat:

Secretary:	J. R. Kline
Associate Secretary:	R. P. Boas
Editorial Committee:	Salomon Bochner (chairman), Einar Hille, P. A. Smith, Oscar Zariski
Financial Committee:	M. H. Stone (chairman), J. L. Coolidge, M. H. Ingraham, John von Neumann, W.L.G. Williams

Organizing Committee: Garrett Birkhoff (chairman), W. T. Martin (vice chairman), G. C. Evans, J. R. Kline, Solomon Lefschetz, Saunders Mac Lane, R.G.D. Richardson, J. L. Synge, Oswald Veblen, J. L. Walsh, D. V. Widder, Norbert Wiener, R. L. Wilder.

The committee stated that it had tried insofar as possible to use the membership of committees as planned for 1940. Deaths, retirements, and personal preferences required some changes. The committee further stated that there was no urgency in designating the President of the Congress and proposed to postpone action. Contrary to the minutes of the Council, the Proceedings of the Congress state that the president designate was chosen in early 1948. The Council approved the recommendations of the committee.

The personnel of the organizing committee changed in several respects prior to the Congress. Richardson died in July 1949. Synge resigned when he left the United States to assume a new position at the Institute for Advanced Studies in Dublin in 1948. Those added to the committee were A. A. Albert, J. L. Doob, T. H. Hildebrandt, Marston Morse, and Hassler Whitney.

The Organizing Committee established four permanent subcommittees and three temporary subcommittees, involving forty-five people. There were subsequent small changes in the committees of the Congress. L. M. Graves was appointed chairman of the Editorial Committee, replacing Bochner, who resigned. A. E. Meder, newly elected treasurer of the Society, was added to the Financial Committee.

A substantial problem facing the Organizing Committee was that of visas for foreign mathematicians who would not ordinarily be eligible under the Ninth Proviso of Section 3 of the Immigration and Naturalization Law of 1918 as amended in 1940. A test case was that of Laurent Schwartz, who had previously been denied a visa. The committee reported that its efforts appeared to be successful. In particular, Professor Schwartz did attend the Congress and gave an address by invitation of the Organizing Committee. Moreover, he received one of the two Fields Medals.

The operation of the Congress of 1950 was rather different from that of the model that has subsequently evolved. For example, the Organizing Committee handled such matters as selecting and inviting hour speakers and appointing a Committee on the Award of the Fields Medals, matters that more recently have been the domain of the International Mathematical Union rather than any local committee.

The Congress was held from 30 September through 6 August 1950. The elected President of the Congress was Oswald Veblen.

The attendance consisted of:

American and Canadian members	1429
American and Canadian associates	539
Members outside U.S. and Canada	271
Associates outside U.S. and Canada	63
	2302

There were also 248 members who in fact did not attend. Thirty-nine countries in addition to the U.S. and Canada supplied members. Every state in the U.S. except South Dakota was represented.

There is a brief report on the Congress in the *Bulletin*, 57 (1951), 1–10. The official report of the Congress consists of two volumes titled *Proceedings of the International Congress of Mathematicians of 1950* and published by the American Mathematical Society in 1952. One is the report of activities and the invited addresses and the other is the contributed papers.

The budget of April 1950 for the Congress called for income of $104,000, with an estimated 1100 members at $15.00 and 500 associates at $7.50, and for expenditures of $101,516. Room and Board for 225 foreigners for nine days at $6.00 per day accounted for an expenditure of $12,150. Not appearing anywhere in this budget is the contribution of Harvard University in services and space.

Inasmuch as the attendance was much larger than the figures used in the budget, there was a substantial surplus. The balance sheet of 30 November 1950 (then the end of the fiscal year) showed that there was cash on hand from the Congress of $45,531.48. Of course there were still outstanding obligations, including the commitment to publish the two volumes of the report. The fund was not kept completely separate. In 1951–1952 there was a deficit in Society operations that was made up by authorization of a transfer from the Congress fund of $13,000. It was further authorized in 1952 to set up a small fund to publicize the Congress of 1954 in Amsterdam and to close out the remainder of the fund, which came to $20,417.59, into surplus.

The terms under which the International Mathematical Union operated had expired before the Congress of 1936, so that the invitation of the United States to hold the 1940 Congress went directly to the 1936 Congress. The 1950 Congress was simply the postponed 1940 Congress and the International Mathematical Union had not been revived.

In 1947 the Society asked the International Council of Scientific Unions to sponsor a meeting of mathematicians in connection with the meeting of UNESCO in Mexico City in November 1947 to discuss plans for the establishment of the International Mathematical Union. When this sponsorship was not forthcoming it was decided to postpone the question until 1950, when the

Congress was in session. A subcommittee consisting of M. H. Stone (chairman), J. R. Kline, and Marston Morse was appointed to plan for a meeting of representative mathematicians just prior to the Congress to discuss the question of the Union. Stone reported at the Council of 1 September 1949 that progress had been made in the appointment of nearly a dozen national committees to consider the problem. At the Council of 29 December 1949, the Policy Committee reported that a subcommittee had been conducting negotiations with national committees of other countries to the end that a conference aimed at formation of a new International Mathematical Union would be held in New York during the three days preceding the Congress of 1950. The meeting was in fact held on 27–29 August 1950 at Columbia University. Statutes and bylaws were adopted. An international ad interim committee was established and the Union was to be declared in existence as soon as ten national committees declared adherence and adopted the statutes and bylaws.

ICM 86

As early as 1980 it was suggested to the Society by the International Congress Office of the U.S. Department of Commerce that the Society invite the International Mathematical Union to locate the International Congress of Mathematicians of 1986 in the United States. After discounting the motivation behind the suggestion, the Executive Committee and Board of Trustees (ECBT) did think that the idea of holding the 1986 Congress in the United States had merit. Observing that it was the place of the U.S. National Committee for Mathematics (USNCM) to initiate such action, Richard S. Palais, chairman of the Board of Trustees, speaking for the ECBT, brought the matter to the USNCM with a favorable recommendation.

In March 1981 the USNCM and representatives of the National Research Council met with representatives of the Association for Symbolic Logic, the Institute of Mathematical Statistics, the Society for Industrial and Applied Mathematics, and the American Mathematical Society.

The following motion was approved:

> Contingent on the acceptance of financial responsibility for the 1986 Congress by the American Mathematical Society, the U.S. National Committee for Mathematics recommends that an invitation be tendered to the International Mathematical Union to hold the 1986 International Congress of Mathematicians in the United States.

A Site Selection Committee was authorized. Its membership was William J. LeVeque, convenor, and H. Hope Daly, both appointed by the USNCM,

Paul T. Batemen, appointed by the AMS, and F. Reese Harvey, appointed by the USNCM.

The Department of Meetings adapted its site selection procedures for joint meetings to the problem of a suitable location for ICM 86. The site inspection at the University of California, Berkeley was conducted in July 1981 by H. Daly and E. Pitcher for the Site Selection Committee with the cooperation of Morris W. Hirsch, incoming chairman of the Department of Mathematics. Following the recommendation of the committee, an invitation was solicited from the University of California and was issued by Chancellor I. M. Heyman. It was accepted by G. D. Mostow for the USNCM. The National Academy of Sciences then issued the invitation to the IMU, which approved in Warsaw in 1982.

From the moment that the holding of the Congress in the United States appeared likely, the secretary was insistent that it be handled by a separate corporation. The Board of Trustees agreed to do this and the ICM 86 corporation was established in 1983 as a Rhode Island non-profit membership corporation with the Trustees of the AMS as its members and five of them as directors along with an executive director. Dr. Jill P. Mesirov, who was assistant executive director of the Society, became executive director of ICM 86 and continued in that capacity after her term as assistant executive director of the Society ended. H. Hope Daly was the Congress manager and Lee Ann Lima the business manager.

There was a Steering Committee consisting of Andrew M. Gleason, chairman, John W. Addison, Jr., Yousef Alavi, Peter J. Bickel, Hirsh G. Cohen, H. Hope Daly, Solomon Feferman, Leon A. Henkin, Shirley A. Hill, Richard M. Karp, Linda Keen, Jill P. Mesirov, Henry O. Pollak, Kenneth A. Ross, Hugo Rossi, and Shmuel Winograd.

The Local Arrangements Committee was made up of John W. Addison, Jr., chairman, Mary Ann Addison, Henry L. Alder, Lenore Blum, William G. Chinn, Ginette Henkin, Irving Kaplansky, Robion C. Kirby, Eugene L. Lawler, Calvin C. Moore, Robert Osserman, P. Emery Thomas, and Anthony J. Tromba.

The Committee on Special Funds was composed of Richard D. Anderson, chairman, Gerald L. Alexanderson, Morton L. Curtis, Susan J. Friedlander, David Gale, Andrew M. Gleason, Daniel Gorenstein, Jill P. Mesirov, K. Brooks Reid, executive secretary, Alice T. Schafer, and Albert R. Stralka.

There was an Advisory Board consisting of George E. Brown, Jr., Edward E. David, Jr., Gerard Debreu, Marvin L. Goldberger, Ira Michael Heyman, Brockway McMillan, I. M. Singer, and James M. Vaughn, Jr.

The Public Information Committee had as its members Donald L. Albers, Yousef Alavi, director of publicity, Gerald L. Alexanderson, advisor, William

G. Chinn, Jane M. Day, Ronald L. Graham, advisor, Peter J. Hilton, Lester H. Lange, and Jean J. Pedersen.

It is not intended to develop an account of the Congress itself beyond this indication of structure and Society involvement. The Program Committee, which invited major speakers, was a committee of the IMU. The acceptance of abstracts and the scheduling was done by Kenneth A. Ross and Hugo Rossi. The president of the Congress was Andrew M. Gleason, who also edited the *Proceedings*.

The Society advanced money for start-up costs of ICM 86 and was reimbursed with payment of some interest. The Society did work for the Congress in the form of sale of services. The Congress had diverse and substantial expenses, including payment to the Society for production of the *Proceedings* and their distribution to members of the Congress. There was income from grants, registration fees, and contributions, including one of $43,000 from the Society. The budget was on the order of one and one-quarter million dollars. When all was over the Congress had debts of about $140,000 and assets consisting chiefly of the copyright to the *Proceedings* and a stock of printed volumes. The Society assumed the debt and ICM 86 turned all of its assets over to the Society. It will not be clear for several years what the Society can realize from the sale of additional copies of the *Proceedings*.

Membership and Dues

Classes of Members

According to the bylaws (VII, 1) there are two kinds of individual members, ordinary and contributing. The contributing member differs from the ordinary member only in voluntarily paying annual dues which are larger by at least 50%. The route to ordinary membership is by application, election, and the payment of dues. In 1938, the election was by the Council. As part of an extensive revision of the bylaws in 1948, an Executive Committee was established and was empowered to elect members. The Council nonetheless continued to exercise the function of election of members, sometimes by mail. It may be the case that the first election by the Executive Committee was at its meeting of March 31, 1962. By 1963, election was being handled by the Executive Committee as business by mail. Since 1983, this duty was further delegated to the secretary and the associate secretaries.

For many years it was required that an application for membership carry the signatures of two members in support. In practice it came about that the signatures were sometimes those of the secretary and of another supplied by the secretary. It appeared that no useful purpose was served by the signatures and in 1973 the practice was abandoned.

Two large classes of ordinary members are nominees and members by reciprocity. Membership by reciprocity allows members of one mathematical organization to become members of a second with reduced (approximately half) dues. Reciprocity had been established with the London Mathematical Society in 1922 and was authorized with other societies in 1930. By 1938, there were reciprocity agreements in place with the Unione Matematica Italiano, the Deutsche Mathematiker-Vereinigung, and the Greek Mathematical Society. The total number of members by reciprocity was small, being 52 in 1937.

Reciprocity was established with the Société Mathématique Suisse in 1941 and with the Sociedad Matemática Mexicana in 1943. In 1946 the Council passed a resolution favoring reciprocity agreements in principle. From this point on, many such agreements were completed, until for the year 1988 there were 48 such agreements in effect. The number of members by reciprocity in 1987 was about 2800.

There was one notable instance of difficulty with membership by reciprocity. Such a membership was established with the South African Mathematical Society in 1972 after an investigation of whether there was any color bar to membership in that Society. It was reported that there was none. In 1974 in response to objections to apartheid the agreement was cancelled.

The philosophical basis for reciprocity is this. There are persons who because of geographic distance or national situation might find that they could not avail themselves readily of the full range of privileges of membership. These privileges include most notably meetings but secondarily a variety of services such as the annual survey and the employment register. Reciprocity is an effort to price the membership for these persons commensurate with the benefits available to them, principally publications, so that they will nonetheless choose to belong.

The issue of possible reciprocity in North America was considered as early as 1950 by the Council. The Council approved a report from a committee consisting of G. T. Whyburn, chairman, Nelson Dunford, R. L. Jeffery, C. C. MacDuffee, and D. C. Spencer, that recommended extending reciprocity agreements "outside the American area." The committee noted that the majority of Canadian mathematicians are members of the AMS and that the Society could not afford the loss of revenue that reciprocity in North America would entail.

At the same time, at the insistence of the Council, the committee stated that "it is strongly recommended that, as occasion permits, mathematicians residing in contiguous territory to the United States be extended cordial invitations to become members of the AMS and that these invitations be supplemented by offering a favorable exchange rate where possible." This addendum was intended to ease the plight of mathematicians in Mexico.

There is a special relationship with the Canadian Mathematical Society, less favorable than reciprocity, established in 1988 whereby a member of the latter may join the Society and the Mathematical Association of America at a slightly reduced rate and a member of the AMS or the MAA may join the CMS at a slightly reduced rate. Note in this respect that the CMS represents the interests of both research and teaching while these interests are somewhat separated between the AMS and MAA. Moreover the CMS and the AMS and MAA have observers with status at the meetings of each other's governing bodies.

The class of institutional member was established in 1934 (see the discussion of dues). Institutional members as a privilege of membership may name nominees, who are ordinary members who do not themselves pay dues. All but a small number (it was usually three) must be graduate students, and may remain nominees for only a short period of time (two years was the established interval). The intent was to interest prospective mathematicians in becoming more permanent members.

There were several other small categories of ordinary membership involving reduced dues for special cases. Of possible interest is the category of life member, in which an individual became a member for life through one payment of a fixed sum. With increasing services offered by the Society and with increased cost of services, partly caused by inflation, it became financially undesirable for the Society to continue this form of membership. The category had been established in 1898, with a payment of $50. More recently the amount was set on an actuarial basis. Life membership was terminated by an amendment to the bylaws in 1941 in that no new life memberships were accepted. In 1988, the only remaining life members in this cohort were John Arnold, Oliver Collins, and David Widder.

In 1986 life membership was reinstated under different terms. A member for the past twenty years and of age at least 62 can become a life member by a single payment of five times the higher level of dues. In 1987 this provision was modified to set realistic dues for life membership of long-time members by reciprocity.

There are several small classes of members related to the dues that they pay. At the meeting of 27 April 1951, a report was received from a Committee to Consider Dues of Members Who Are the Wives of Members consisting of C. R. Adams, Chairman, E. R. Lorch, and D. V. Widder. The principle was enunciated that "Members of the Society ought normally to think in terms of their membership and their dues as support for a worthy enterprise in which they have a special interest, with relatively small emphasis on what they receive." There were several considerations in the remainder of the report but the matter was tabled.

In the minutes of the Executive Committee and Board of Trustees of 5–6 June 1965 there is the following item:

> The establishment of $28 as membership dues for a husband-wife joint memberships [sic]. The husband is to be billed at the rate of $28 for dues and will receive the *Notices* and *Bulletin* as a privilege of membership. The wife will pay no dues but will be allowed a choice of subscriptions at members' rates, and both will be accorded all other privileges of membership.

It should be observed that the ordinary dues at the time were $20 and the members' rate for the *Notices* and *Bulletin* was $12, so that the figure of $28 paid for two memberships at $20 less one *Notices* and *Bulletin*.

A similar formula in less sexist language has prevailed. The following arrangement is available to a couple choosing to use it. One spouse pays full dues at the lower or higher level based on professional income. The other pays dues based on professional income less $20 and does not receive the *Notices* and *Bulletin*. The figure $20 is a vestige of a time when this was the members' rate for the *Notices* and *Bulletin*.

There is a category called student membership. A nominee, who has been paying no dues, may find full dues a hardship and drop membership. The category of student member was established effective in 1972 for discontinued nominees, good for one year whether the individual was in fact a student or not. The dues figure was set at $10, then half of the regular dues. It has gone up to $12. This category was extended to full-time students and unemployed (not resigned or retired members) without limit of time effective in 1974. The only verification of status is a statement signed by the member.

Members of at least twenty years who retire because of age or disability may remain active members without payment of dues. This level of membership however does not include the *Bulletin* as a privilege.

The number of members in a given year depends greatly on the day in which the count is made. The most abrupt change takes place in the late spring when persons who did not renew for the current year are dropped from membership. For many years, the count has been based on the Combined Membership List (AMS-MAA-SIAM), which is compiled during the summer and is printed in the fall. The recorded numbers of members in several years are as follows:

Year	Number
1940	2314
1946	3097
1950	4386
1955	4878
1960	6725
1965	10923
1970	14197
1975	15907
1980	19984
1985	22031
1987	22611

The more recent numbers are slightly inflated. They are the numbers of names of Society members in the Combined Membership List. There is a policy that when a person is dropped for nonpayment of dues, say about 1 May, the name still appears in the CML being compiled. This policy insures that names and addresses are not lost to the mathematical public through inadvertent failure to pay dues.

In his report of activity of September to November 1955 Executive Director John Curtiss advanced the idea of selling membership buttons as a morale builder and to increase group consciousness. He stated that such an emblem had been successful in other societies and outside the academic scene. The Council of 28 December 1955 empowered the executive director to explore the idea. Following a survey of a sample of membership the Council authorized the button, which carried the design of the seal, much as it appears on the cover of the Notices and elsewhere, and the Trustees concurred. Later the emblem was authorized for a tie clasp as well.

Some dissatisfaction with the idea of the emblem was expressed and the Council of 24 April 1959 followed a recommendation of the secretary that the emblem no longer be advertised though any orders that materialized could be filled. Some time later sales were stopped completely.

Election to membership has seemed to be almost automatic. For many years, persons proposed (i.e., applying) for membership were listed in the minutes for one meeting and were elected at the next. However there is a case recorded in the minutes of 23 November 1945 of an individual proposed for membership and not elected. There is no indication of reason and no record of the application.

More interesting is the case of Bourbaki, the faceless consortium of mathematicians concerned with definitive exposition. On 12 November 1948, an application for membership was received from Nicolas Bourbaki. According to this, he was born in Cucutemi, Poldavia on 29 February 1885 and received his Ph.D. in 1906 from the Royal Poldavian University. He was currently a fellow of the Rockefeller Foundation, having been a professor at the Royal Poldavian University from 1910 to 1919, scientific adviser to Hermann Publishing Co. since 1934, and member of the Royal Poldavian Academy since 1914. The recommendation for election came from A. A. Albert and L. M. Graves.

On 8 January 1950 another application was received from Nicolas Bourbaki. This time he was born in Cucutemi in the province of Moldavia in Rumania on 12 December 1886. His degree was Doctor in Mathematics from Kharkov University in 1910. He held the position of Directeur libre de Recherches a l'University [*sic*] de Nancy with a street address in Nancy. He had been Privat Dozent at the Dorpet University in 1913–1916 and professor at Zorngale College in 1926–1929. He was a member of the Royal Academy of Poldavia since 1917 and of the French Mathematical Society since 1949. This application was supported by R. M. Thrall and T. H. Hildebrandt. There is a pencil notation "à remplir et à retourner à FOURÈS." Subsequent correspondence indicates that Thrall had received the application from Professor Fourès. (L. Fourès handled the applications for membership by reciprocity for the Société Mathématique de France and Thrall received them for the AMS.)

The secretary, J. R. Kline, had rejected the first application, saying to Albert in part that "The Society has two types of membership, individual and institutional contributing memberships. Bourbaki comes under neither classification." No more was heard from this application.

With respect to the second application, Kline wrote to Hildebrandt, a recent past president with whom he had worked, explaining the facts and giving a detailed comparison of the two applications. He further stated "I realize, of course, that you signed this application without having the background. I do not like this type of transactions [*sic*]. In fact, I rather resent membership in the Society being made a matter of jest. I think all of us who are devoted to and have worked for the Society believe that its dignity is such that one should not consider membership in it so lightly." He inquired how to write to Thrall. Hildebrandt replied that "I played the role of the innocent bystander or shall we call it dupe." He further stated that Thrall brought him the application, saying that it was from a real person and that he had checked the places mentioned and they seemed genuine. He suggested a letter to Thrall similar to the one he had received, which Kline then wrote. The tone of Thrall's reply suggests that indeed he did not know the background. On the other hand, he noted that "since N. Bourbaki is a member of record of the French Mathematics Society the matter of our rejecting his application carries implications which may be a source of international complications." He further suggested that the matter be considered by the Council. Although he represented the Société Mathématique de France for the AMS in matters of reciprocity, he was unwilling to pursue the matter.

Kline wrote a long letter of explanation to past president Einar Hille suggesting a mail vote of the Executive Committee. Another consideration appears in the letter, namely the potential of boycott of the incipient International Congress of 1950 in the United States were feathers to be ruffled. He then wrote to Saunders Mac Lane, W. T. Martin, and J. L. Walsh, all members of the Executive Committee and the last the president, seeking advice. He also wrote to Warren Weaver asking whether Nicolas Bourbaki was a Fellow of the Rockefeller Foundation on October 27, 1948.

Walsh noted that any decision was likely to be wrong, that the decision must be made by the Council, and that this is conventionally a slow process, which should not be hurried in this instance.

Hille replied in part that "I am afraid that I belong to the people who will never grow up, but it seems to me that you are running the risk of making a mountain out of molehill.... It may be a joke, but may also be for serious reasons to get the benefit of membership. In either case I would advice[*sic*] against a formalistic attitude." Further on, he wrote "I think a good case can be made out for granting membership. After all, a good fictitious character lives more intensely and a good deal longer than the humdrum reality. We know much more about Ulysses, who probably never existed, than about

Thales who did. There is no question in my mind, but that N. Bourbaki has made a stronger impression on present day mathematics and his fame will last longer than that of most present members of our Society...."

Martin recommended rejection, thought it was improper that the application had been filed, and recommended the formal vote of the Executive Committee, with the possibility of taking it to the Council.

Mac Lane remarked that "The whole matter has a definitely *humorous* side — why can't it be stalled along and brought to the attention of the Council for their amusement?"

Weaver stated that "The Rockefeller Foundation has never given a fellowship or any other grant to an individual mathematician whose name was or is Nicholas [*sic*] Bourbaki. We did make a grant in aid to assist the activities of the Bourbaki group. Our grant was made to the University of Nancy, and our action refers to the 'Bourbaki group.' " He went on to say "I would not myself be inclined to be peeved about the matter, but would rather write to some responsible member of the group, such as André Weil, and draw to his attention the regulations of the Society which provide for either individual or institutional membership only. If the Bourbaki Group wish to present their corporate entity, call it what they will, for membership, this obviously should be done in a serious and responsible way."

The issue did come before the Council of 28 December 1950, where a motion was passed that "the proper persons be notified that (1) Bourbaki cannot be elected to individual membership in the Society under the reciprocity agreement between the French Mathematical Society and the American Mathematical Society, (2) Bourbaki may be elected to institutional membership at the rate of $25, per year, and (3) any of the constituents of Bourbaki may become an individual member of the American Mathematical Society."

It fell to E. G. Begle, successor to Kline as secretary, to write to Fourès on 29 February 1951 explaining the Council action. The reply came from J. Dieudonné. The second paragraph says "If the French Mathematical Society took itself as seriously as seems to be the case with the A.M.S., this letter, and the breach of the reciprocity agreement implied therein, could seriously jeopardize the good relations between the two Societies. Professor Nicolas Bourbaki was admitted by the Société Mathématique de France as an individual member in 1949, and has applied in this capacity for membership in the A.M.S.; it is the clear meaning of the first paragraph of the reciprocity agreement that such an application was in order, and that the A.M.S. had no right to scrutinize it any further." He continued in a lighter vein about having referred the matter of Prof. Bourbaki. In particular he writes "I should like to point out that it often happens in the career of a famous mathematician that he gradually loses his individual character and becomes an institution. ... Prof. Bourbaki would of course raise no objection against sharing the fate of some of his distinguished American Colleagues. Pending such a step, I need not tell you that his agreeing to that status is entirely out of the question. As to the group of his friends and collaborators, it is of an entirely informal nature, and is not to be institutionalized in any way." The file ends at this point.

DUES

The dues of the Society have generally been set at a level to cover cost of the services offered. In periods during which there were few increases in service and little inflation it was possible to keep the dues fixed for long periods of time. When it became necessary to raise dues, the percent raise was substantial in order to allow the new rate again to remain constant.

The annual dues, set at $5.00 in 1891, were raised to $6.00 effective in 1921 and to $8.00 in 1931.

World War II caused substantial disruption in the affairs of the Society. With respect to dues, enlisted men in the armed forces of the United States and Canada who were members in good standing were granted nominal dues of $1.00. The number of persons availing themselves of this privilege by April 1942 was only 10. The number granted the privilege in 1945 was 35.

The dues were raised to $10.00 effective in 1948 and to $14.00 effective in 1951. It should be noted that even with the dues of $8.00 there was a reduced figure of $6.00 during the first two years of membership and similar arrangements were made with the dues levels of $10.00 and $14.00.

Up to 1965, the dues had been established in the bylaws and raised by amendment. In 1965, two changes were made. The dues were raised to $20.00 effective in 1966. Then the bylaws were amended to lodge the setting of the figures for dues with the Council, with the approval of the Trustees. In practice, subsequent suggestions of dues increases, even including informal advance approval, came from the Trustees.

With the next increase in dues, a new principle was introduced. Dues were set for 1974 at $32.00($24.00) with the understanding that persons with annual professional income under a cutoff of $15,000 should pay the lower figure stated in parentheses. The figure was raised to $48.00($36.00) effective in 1979. At the same time the cutoff was raised to $20,000.

A simplified procedure for handling dues increases was established by the Council in January 1982. The effect is to have small annual increases rather than large increases at longer intervals. It was agreed that the Council establish a formula by which dues are increased each year a whole dollar amount equal to the percentage increase in the average salary for institutions reporting data to the American Association of University Professors for the last decade, that the cutoff be at a round thousand dollar amount placing about

60% of the membership at the lower dues level, and that the Executive Committee be empowered to carry out these steps for the Council.

The AAUP figure was used rather than the Consumer Price Index because of widespread criticism of the way the latter incorporates the cost of purchase of housing into the index and because the AAUP index was thought better to suit the circumstances of a majority of the members. The procedure was introduced as less cumbersome than ad hoc setting of dues in the Council and as more adaptable to making changes effective in a reasonable span of time. The setting of dues can take place at a single joint meeting of the Executive Committee and Board of Trustees in May to be effective the following January.

When the Society ran a substantial deficit in 1983 it was recognized that although the scheduled annual increase was reasonable the base from which it began was too low. Dues were raised sharply for 1985 and 1986 to increase the base. The formula was then put in place again for 1987 and beyond.

The Executive Committee and Board of Trustees of November 1987 concluded that the increase of 1985 and 1986 may have been too large. They recommended to the Council that the dues not be increased for 1989 though the formula continue to be applied thereafter. The Council concurred for 1989 and left the formula in place for 1990 without foreclosing action for 1990.

With this system in place, dues have proceeded as follows:

Year	Dues	Cutoff
1983	$52.00(40.00)	$21,000
1984	56.00(44.00)	21,000
1985	66.00(50.00)	23,000
1986	78.00(59.00)	24,000
1987	84.00(64.00)	26,000
1988	88.00(66.00)	30,000
1989	88.00(66.00)	38,000

Some trivial comments should be made. The lower dues were intended to be three-quarters of the higher dues. The figure 40 rather than 39 occurs in 1983 because it is convenient that the number be even in that half of it occurs elsewhere in the dues schedule. Note however that 59 occurs in 1986. The number 44 in 1984 is the result of a programming error. The program computed 40 from 52 in 1983 by subtracting 12 rather than by multiplying by .75. The same step was performed in 1984. It has been observed in retrospect that the cutoff has tended to be too low.

Dues of Institutional Members

Institutional membership, already mentioned in connection with nominees, is a plan by which an accredited educational institution in North America can become a member of the Society for a fee whose size depends on the amount of research activity at the institution. As early as 1933–1934, a plan for institutional contributing members was put in place. The initial effort, involving visits by Professor Mark H. Ingraham, associate secretary for Financial Affairs, to 85 institutions to explain the plan, is described in [A].

In December 1940, Ingraham reported on the first six years of the plan. In 1940 there were 49 institutions paying $25 in dues and 34 paying somewhat more, for a total income of $6,205.00. (Because of a change in accounting there may be a few institutions that paid twice during the reporting period, so that the total number of institutions is less than 83.)

Initially the bylaws made provision for institutional contributing members and set minimum dues of $25.00. Dues were actually set according to the amount of research published by the faculty in the *Bulletin, Transactions, American Journal*, and the *Annals*. The bylaws were substantially revised in the interval 1948–1950 and two versions of institutional membership appear there. In 1949 the dues and privileges are enumerated. However in 1950 the current form of the bylaw appeared as follows:

> The minimum dues of an institutional member shall depend on the amount of published material credited to that member in certain journals during a specific period. The formula for computing these dues shall be established from time to time by the Council, subject to approval by the Board of Trustees. Institutions may pay larger dues than the computed minimum.

A principal part of the process was a formula that consisted in counting pages published in an ever increasing list of research journals over a three year period and multiplying by a dollar-per-page factor. By the 1980s, the list of journals had grown to 63. The price per page fluctuated at $1.00 to $1.50 per page. Also the dues of the previous year were multiplied by an inflation factor. The larger of the two numbers thus obtained, rounded to a multiple of $25.00, was the dues figure. Note the effect that under this formula dues never decrease.

Institutional membership allowed a substantial discount on publications, included several "free" publications such as *Bulletin, Notices, Combined Membership List*, and *Professional Directory* (formerly *Administrative Directory*), and as already noted allowed the naming of nominees, who received free membership. The number of the latter was set at three for minimum dues and increased with larger dues. Except for three, the nominees were to be graduate students.

The membership is a great bargain for an institution at minimum dues that wishes publications, particularly *Mathematical Reviews.* The difference between list price and institutional members' price of the last alone exceeds minimum dues. Of course the membership encourages the purchase of some items that might otherwise go unsold. The membership is a modest bargain for larger and more active institutions with the exception of a handful in that they at least break even with the price of publications and have the advantage of being able to offer membership to their graduate students. The latter is a great advantage to the Society in securing members in that some nominees continue as paying members.

For a handful of institutions, the page count produced a figure large enough to be uneconomic. Almost all such institutions continued to pay but there was grumbling and threat of defection so that the dues at the highest level were reduced. In 1987 a Trustees' Committee on Institutional Membership, consisting of Frederick W. Gehring, chairman, Ramesh Gangolli, William A. Veech, James A. Voytuk, ex officio, and Carol-Ann Blackwood, consultant, proposed that the base of calculation of institutional dues be changed from "the amount of published material credited to that member" to the amount of scholarly activity, measured in terms of published research, size of faculty, and number of graduate students. The Council of 5 January 1988 approved the principle and as this is written a suitable formula is being devised.

The "sunshine laws" of Florida (and potentially of other states) cause a problem with respect to institutional membership. According to some versions of a sunshine law the state requires a statement from an organization that all its records are open to inspection as a condition for paying membership dues of one of its subdivisions or arms to the organization. On the other hand, it is clear that the Society cannot supply such a statement inasmuch as its records involve reports of referees. Compliance with the statement would open these reports to fishing expeditions by disgruntled authors. The absence of the statement prevents state universities in such states from holding institutional membership. To obviate this difficulty, the Society tailors and prices a package of journals to meet the needs of an institution that would be eligible for institutional membership were it not for the legal barrier. This arrangement goes under the name "temporary unified subscription."

The Council

Contemplation of the Council over a period of fifty years reveals changes in composition, voting strength, and function.

The Council consists of members-at-large and ex officio members. The number of members-at-large, called elected members in earlier times, has been fifteen since 1924. The changes in composition and in voting strength in the Council with time have taken place among the ex officio members.

In 1938 the Council consisted of the president, three vice presidents, the secretary, four associate secretaries, the treasurer, the librarian, four editorial committees of three persons for the *Bulletin,* the *Transactions,* the *Colloquium Publications,* the representatives on the *American Journal,* any ex-secretary who had served for at least ten years, ex-presidents for six years after their terms as president, and fifteen member-at-large.

Certain other named committees have been represented on the Council by their full membership. When *Mathematical Reviews* began in 1940 and *Mathematical Surveys* and *Monographs* in 1944, these were added. When the *Proceedings* was split from the *Bulletin* in 1950 and when *Mathematics of Computation* became an official Society journal in 1966, these were added. The members of the Committee on Printing and Publishing, which was concerned with non-editorial matters and existed in the interval 1953 to 1971, were Council members.

Council membership by members of editorial committees was changed in 1985 in a manner to be described as part of the account of voting rights.

The Committee to Monitor Problems in Communication, which had been a standing committee, gained representation by its chairman in 1971. The Committee on Science Policy, which had been a standing committee, was represented by its chairman since 1985.

Up to 1938, the Nominating Committee traditionally offered one candidate for each position. In 1940 there were two petitions that the bylaws be revised, in particular to allow for freer methods of nominating and electing members of the Council. A committee consisting of L. P. Eisenhart, chairman, C. R. Adams, W. C. Graustein, W. T. Martin, and A. D. Michal, made a proposal, the burden of which was that the membership should elect members-at-large in contested elections and that the Council elect officers and editors. The chairman himself opposed the recommendations in a covering letter. There was no action on this formulation. A committee consisting of W. M. Whyburn, chairman, A. Church, and P. Franklin produced a revision that was approved by the Council and the membership in 1941.

The revision added the Editorial Committee of *Mathematical Reviews* to the list of editorial committees and did not otherwise change the composition of the Council. However, it changed the voting strength in that each editorial committee designated one voting member (who might be replaced by an alternate). Associate secretaries (unless one was acting secretary) did not vote. Moreover, although the bylaws did not so specify, elections for vice-president and member-at-large were contested. From that point on, the majority of the voting strength lay with members chosen in contested elections.

The total result was not completely satisfactory. A Committee on Reorganization, consisting of W. L. Ayres, chairman, S. S. Cairns, B. P. Gill, J. R. Kline, W. T. Martin, P. A. Smith, and M. H. Stone made several recommendations to the Council of 16 April 1948. These included the creation of the office of president elect, the creation of an Executive Committee with delegated powers, and some changes in the Council. The difficulty with the Council lay in the editors and associate secretaries, who were members without vote. The proposed solution was that all editors should be voting members but that the set of four associate secretaries and the editorial committees should each have one vote, divided among the members present. Matters were to be decided by simple majority vote but in the event that a count was requested it was to be carried out "by fractions."

The Council accepted these recommendations and turned the writing over to a committee consisting of S. S. Cairns, chairman, J. R. Kline, and P. A. Smith, who produced the required draft at the Council of 30 August 1949. It was approved by the Council and the membership. In particular the term of an ex-president was limited to one year and of an ex-secretary to two years.

The position of librarian disappeared in 1952 when the library was sold. See the section on the move from New York to Providence.

The position of associate treasurer as a council member appeared in 1973.

By 1985 the Council as described above had become too large. The possible maximum of 41 in 1938 had grown to 66, mostly with the increase in number and size of editorial committees. Moreover, the voting strength had changed in a subtle and unintended manner. There were fifteen members at large and three vice-presidents, all elected in contested elections. To maintain the principle that persons so elected should be in the majority, single votes had been assigned to each editorial committee and to the set of associate secretaries, with voting done by voice vote unless there was a request for a roll call, when it was done "by fractions" as already described. As the Council grew larger, there were more voices, both in the debate and in the voice votes, from persons chosen in uncontested elections even though the distribution of votes by fractions remained unchanged.

The Executive Committee proposed that only the chairman of an editorial committee (or a deputy) be a Council member, that only the associate secretary for the scientific program (or a suitable alternate) be a member and, incidentally, that the chairman of the Committee on Science Policy be made a member. The need for fractional votes would vanish. The Long Range Planning Committee endorsed this position. The Council of 8 January 1985 approved the amendment to the bylaws and so did the Business Meeting of 14 August 1985. The expected clause that no one already on the Council on 1 January 1986 should be displaced during the current term was included. The effect was to reduce the size of the Council over a four year period to a maximum of 39, achieved only when there is an eligible past secretary and when three of the four elected members of the Executive Committee have had their terms extended to allow for that duty. The voting by fractions was eliminated.

NOMINATIONS

The bylaws specify that nominations are made by the Council. In practice, a Nominating Committee brings recommendations to the Council, which almost invariably follows them. With respect to editors, the Council of 29 December 1938 instructed the Nominating Committee to consult the appropriate editorial committee in making nominations. Recently proposals for vice-president and member-at-large have also come to the Council by petition and, except in the case of repeated petitions for the same candidate, these are followed also.

It has already been noted that the nominating procedure was liberalized in 1941 in that the election for member-at-large and for vice-president was contested.

For many years the Nominating Committee was named by the President, who usually chose a new committee each year with the possible exception of holding one person from a previous year to serve as chairman. Two objections appeared. One was that it perpetuated an "old boy" effect. The other was that, inasmuch as the election of a president is uncontested, each president is naming his successor at one remove.

At the meeting of 31 August 1971, the Council agreed to hear a presentation by Mary Gray in which she proposed that suggested candidates supported by petition be nominated. The Council agreed to consider such a plan with respect to candidates for member-at-large and vice-president. A committee consisting of M. Gerstenhaber, chairman, P. R. Halmos, and V. L. Klee,

Jr., presented a motion on the proposed lines at the Council of 16 January 1972. The number of signatures on a valid petition was set at 50 after discussion. The most intense argument came over the issue, raised by Halmos in a minority opinion, who wrote as follows:

> The ballot ... nominates people who are to help make decisions of scientific import; I think a certain quantity of proved scientific competence should be a prerequisite .... I prefer the ... procedure of nominating deserving and competent people by good will and good sense to one legalistically circumscribed. If, however, that procedure is called into question, ... then I urge a very carefully legalistically formulated set of qualifying conditions that a candidate must satisfy.

After extended argument an amendment by Saunders Mac Lane was passed to include a qualification that the candidate must have had five books or papers reviewed in *Mathematical Reviews* and that the petitioners should present a statement on the mathematical accomplishments of the candidate. In this form the motion was returned to the committee.

The committee, reduced to two with the resignation of Halmos, presented a revised resolution that did not contain reference to scientific qualifications. The Council made changes in details and authorized a two-year trial.

Nomination by petition (that is, proposals for nomination) became a permanent feature of the nomination procedure. In the initial year 1972, there were seven candidates by petition, of whom one, Anatole Beck, was elected. In subsequent years through 1987 there appear to have been 17 such candidates of whom three were elected. In the same spirit, effective with the Nominating Committee of 1976, the committee was elected by the membership in a contested election from candidates named by the president together with candidates proposed by petition, if any.

In the election to the Nominating Committee, the "single transferrable ballot" is used. Opinions vary as to the utility and effectiveness of this method. The secretary has served as teller in these elections and has observed that in filling four places there have been two occasions in thirteen years when the person scoring fifth in first choice votes has been elected through the effect of the transfers.

There has been some objection to having more than one member-at-large from an institution. Recommendations of the nominating committee that violate this convention have been rejected, as in 1954 when two names from the University of Illinois were suggested. More recently, the nominating committee has been advised to consider seriously any proposed nomination that places more than three Council members or one Trustee at an institution. With the reduced size of the Council this advice can be expected to change.

At the time of preparation of this book arrangements were being made to have editorial commmittees named in the bylaws appointed by the Council on the basis of nominations made by an elected Editorial Boards Committee (EBC). The principle had been established but not all of the details of mechanism. The EBC was also to be charged with monitoring the functioning of those editorial committees.

Council Members

There is a separate chapter devoted to the presidents of the second fifty years. There is a chapter for the other officers, namely vice-president, secretary, associate secretary, treasurer, associate treasurer, and librarian.

The members of the editorial committees are listed with the accounts of the publications that they edited. Those elected before 1 January 1986 were Council members during the entire term covered by the election. One elected after that date was not a Council member unless serving as chairman of the editorial committee. (The title was sometimes managing editor.) Members from the Committee on Printing and Publishing, the Committee to Monitor Problems in Communication, and the Committee on Science Policy are named in the discussion of those committees.

It is only in recent years, beginning in 1965, that travel costs of Council members to Council meetings have been paid. These are limited to economy plane fare or the equivalent and some ground transportation.

Members-at-large

Neither members-at-large nor ex officio members are eligible for immediate election as member-at-large. There is, however, no bar to election more than once, though this rarely happens. Fourteen persons have served as member-at-large for two terms. A member-at-large may simultaneously be an editor or may follow membership at large by membership in some other capacity. There are persons who have been members of the Council almost continuously over many years. The prime example is P. R. Halmos, with four terms as member-at-large and one as vice-president, as well as membership on the Council on editorial committees and the Executive Committee.

The terms stated in the following list are the statutory terms of three years. A member of the Council elected to the Executive Committee for a term extending beyond the statutory term remains on the Council and serves as a member-at-large. No account has been taken of that phenomenon in this list. It is not certain that all irregularities such as resignations have been flagged. Here are the members-at-large of the second fifty years:

W. L. Ayres	1/37–12/39	I. S. Sokolnikoff	1/45–12/47
G. A. Hedlund	1/37–12/39	Ralph P. Boas	1/46–12/48
H. L. Rietz	1/37–12/39	R. H. Cameron	1/46–12/48
Warren Weaver	1/37–12/39	R. V. Churchill	1/46–12/48
G. T. Whyburn	1/37–12/39	Churchill Eisenhart	1/46–12/48
Philip Franklin	1/38–12/40	A. P. Morse	1/46–12/48
A. D. Michal	1/38–12/40	H. W. Bode	1/47–12/49
J. L. Synge	1/38–12/40	Samuel Eilenberg	1/47–12/49
W. J. Trjitzinsky	1/38–12/40	B. O. Koopman	1/47–12/49
John von Neumann	1/38–12/40	Gordon Pall	1/47–12/49
H. E. Bray	1/39–12/41	J. W. Tukey	1/47–12/49
Alonzo Church	1/39–12/41	J. L. Doob	1/48–12/50
E. J. McShane	1/39–12/41	R. L. Jeffery	1/48–12/50
H. P. Robertson	1/39–12/41	J. B. Rosser	1/48–12/50
Hermann Weyl	1/39–12/41	Norman E. Steenrod	1/48–12/50
L. E. Dickson	1/40–12/42	Alfred Tarski	1/48–12/50
C. G. Latimer	1/40–12/42	Emil Artin	1/49–12/51
Saunders Mac Lane	1/40–12/42	Herbert Busemann	1/49–12/51
N. H. McCoy	1/40–12/42	George W. Mackey	1/49–12/51
W. M. Whyburn	1/40–12/42	M. H. Heins	1/49–12/52
L. R. Ford	1/41–12/43	L. C. Young	1/49–12/51
G. B. Price	1/41–12/43	Warren Ambrose	1/50–12/52
Tibor Radó	1/41–12/43	Paul R. Halmos	1/50–12/52
A. W. Tucker	1/41–12/43	Mark Kac	1/50–12/52
Morgan Ward	1/41–12/43	S. B. Myers	1/50–12/52
G. A. Bliss	1/42–12/44	D. C. Spencer	1/50–12/52
Nelson Dunford	1/42–12/44	David Blackwell	1/51–12/53
W. T. Martin	1/42–12/44	H. F. Bohnenblust	1/51–12/53
Gabor Szegö	1/42–12/44	Irving Kaplansky	1/51–12/53
S. S. Wilks	1/42–12/44	H. S. MacDonald Coxeter	
R. P. Agnew	1/43–12/45		1/51–12/53
Harry Bateman	1/43–12/45	W. T. Reid	1/51–12/53
E. T. Bell	1/43–12/45	L. V. Ahlfors	1/52–12/54
Richard Courant	1/43–12/45	C. B. Allendoerfer	1/52–12/54
D. H. Lehmer	1/43–12/45	R. H. Bing	1/52–12/54
Garrett Birkhoff	1/44–12/46	E. R. Lorch	1/52–12/54
M. R. Hestenes	1/44–12/46	J. C. Oxtoby	1/52–12/54
Harold Hotelling	1/44–12/46	F. B. Jones	1/53–12/55
Nathan Jacobson	1/44–12/46	Edwin E. Moise	1/53–12/55
R. D. James	1/44–12/46	B. J. Pettis	1/53–12/55
H. F. Bohnenblust	1/45–12/47	R. M. Thrall	1/53–12/55
S. S. Cairns	1/45–12/47	George W. Whitehead	1/53–12/55
H. B. Curry	1/45–12/47	R. P. Dilworth	1/54–12/56
M. H. Ingraham	1/45–12/47	R. D. James	1/54–12/56

C. C. MacDuffee	1/54–12/56	R. C. Lyndon	1/62–12/64
D. V. Widder	1/54–12/56	Lawrence Markus	1/62–12/64
Antoni Zygmund	1/54–12/57	D. A. Darling	1/63–12/65
E. F. Beckenbach	1/55–12/57	Karel de Leeuw	1/63–12/65
David Blackwell	1/55–12/57	G. F. D. Duff	1/63–12/65
R. H. Fox	1/55–12/57	Louis Nirenberg	1/63–12/65
Kurt O. Friedrichs	1/55–12/57	Leo Zippin	1/63–12/65
G. de B. Robinson	1/55–12/57	Eugenio Calabi	1/64–12/66
John L. Kelley	1/56–12/58	Victor L. Klee, Jr.	1/64–12/66
Hans Samelson	1/56–12/58	John W. Milnor	1/64–12/66
Abraham H. Taub	1/56–12/58	A. E. Taylor	1/64–12/66
A. D. Wallace	1/56–12/58	John Wermer	1/64–12/66
Einar Hille	1/56–12/59	R. F. Arens	1/65–12/67
R. E. Bellman	1/57–12/59	Maurice Auslander	1/65–12/67
Lipman Bers	1/57–12/59	Alberto P. Calderón	1/65–12/67
Andrew M. Gleason	1/57–12/60	Lamberto Cesari	1/65–12/67
Paul R. Halmos	1/57–12/60	Paul J. Cohen	1/65–12/67
D. H. Lehmer	1/57–12/59	J. B. Diaz	1/66–12/68
R H Bing	1/58–12/60	Walter Feit	1/66–12/68
L. H. Loomis	1/58–12/60	Henry Helson	1/66–12/68
P. C. Rosenbloom	1/58–12/60	Kenneth M. Hoffman	1/66–12/68
Walter Rudin	1/58–12/60	Ivan Niven	1/66–12/68
I. E. Segal	1/58–12/60	Richard D. Anderson	1/67–12/69
R. H. Bruck	1/59–12/61	William Browder	1/67–12/69
P. R. Garabedian	1/59–12/61	Avner Friedman[1]	1/67–12/69
Edwin Hewitt	1/59–12/61	Jürgen K. Moser	1/67–12/69
G. de B. Robinson	1/59–12/61	Gian-Carlo Rota	1/67–12/69
Leo Zippin	1/59–12/61	Armand Borel	1/68–12/70
R. Creighton Buck	1/60–12/62	Raoul H. Bott	1/68–12/70
M. M. Day	1/60–12/63	Israel N. Herstein	1/68–12/70
Arthur Erdélyi	1/60–12/62	Serge Lang	1/68–12/70
G. E. Forsythe	1/60–12/62	Jacob T. Schwartz	1/68–12/70
Edwin H. Spanier	1/60–12/62	Hyman Bass	1/69–12/71
Paul T. Bateman	1/61–12/63	Paul J. Cohen	1/69–12/71
H. W. Bode	1/61–12/63	J. Wallace Givens, Jr.	1/69–12/71
Raoul H. Bott	1/61–12/63	Victor L. Klee, Jr.	1/69–12/71
R. S. Phillips	1/61–12/63	Ray A. Kunze	1/69–12/71
C. B. Tompkins	1/61–12/63	Michael Artin	1/70–12/72
Leon A. Henkin	1/62–12/64	Michael F. Atiyah	1/70–12/72
Alston S. Householder		Philip Hartman	1/70–12/72
	1/62–12/64	C. C. Lin	1/70–12/72
Peter D. Lax	1/62–12/64	Calvin C. Moore	1/70–12/72

[1]Friedman was elected by the Council to fill a vacancy caused by the fact that Sigurdur Halgason declined to serve.

George F. Carrier	1/71–12/73
Morton L. Curtis	1/71–12/73
Mary Ellen Rudin	1/71–12/73
James B. Serrin	1/71–12/73
Elias M. Stein	1/71–12/73
William Browder	1/72–12/74
P. S. Mostert	1/72–12/74
Robert T. Seeley	1/72–12/74
Dorothy Maharam Stone	1/72–12/74
Olga Taussky-Todd	1/72–12/74
Anatole Beck	1/73–12/75
Michael Golomb	1/73–12/75
Mary W. Gray	1/73–12/75
Arthur P. Mattuck	1/73–12/75
Cathleen S. Morawetz	1/73–12/75
Charles W. Curtis	1/74–12/76
Herbert G. Keller	1/74–12/76
Robion C. Kirby	1/74–12/77
Lee Lorch	1/74–12/76
Jane Cronin Scanlon	1/74–12/76
David Gale	1/75–12/77
Judy Green	1/75–12/77
Phillip A. Griffiths	1/75–12/77
Karl K. Norton	1/75–12/77
J. Ernest Wilkins, Jr.	1/75–12/77
Hugo Rossi	1/76–12/78
Lee A. Rubel	1/76–12/78
Barry Simon	1/76–12/78
Daniel Waterman	1/76–12/78
Guido L. Weiss	1/76–12/78
Theodore W. Gamelin	1/77–12/79
Richard J. Griego	1/77–12/79
Karl H. Hofmann	1/77–12/79
Henry P. McKean	1/77–12/79
Linda Preiss Rothschild	1/77–12/79
Isadore M. Singer	1/78–12/78
Joan S. Birman	1/78–12/80
Lenore Blum	1/78–12/80
James A. Donaldson	1/78–12/80
Murray Gerstenhaber	1/78–12/80
Ronald L. Graham	1/78–12/81
Stephen S. Shatz	1/79–12/79
Chandler Davis	1/79–12/81
Robert P. Gilbert	1/79–12/81
Johan H. B. Kemperman	1/79–12/81
Karen Uhlenbeck	1/79–12/81
Daniel H. Wagner	1/79–12/81
Frederick W. Gehring	1/80–12/82
Lee Lorch	1/80–12/82
Richard S. Millman	1/80–12/82
Marian B. Pour-El	1/80–12/82
Mary Ellen Rudin	1/80–12/82
Donald L. Burkholder	1/81–12/83
Alan J. Hoffman	1/81–12/83
Linda Keen	1/81–12/83
Paul J. Sally, Jr.	1/81–12/83
David A. Sanchez	1/82–12/82
Isadore M. Singer	1/82–12/82
Peter A. Fillmore	1/82–12/84
Robert P. Langlands	1/82–12/84
M. Susan Montgomery	1/82–12/84
Hector J. Sussman	1/82–12/84
Melvin Hochster	1/82–12/86
Peter L. Duren	1/83–12/85
Susan J. Friedlander	1/83–12/85
Paul R. Halmos	1/83–12/85
Robin Hartshorne	1/83–12/85
Michael Shub	1/83–12/85
Olga Taussky-Todd	1/83–12/85
Michael G. Crandall	1/84–12/86
David Eisenbud	1/84–12/86
Carlos Kenig	1/84–12/86
Jean Taylor	1/84–12/86
William P. Thurston	1/84–12/86
A. T. Bharucha-Reid[2]	1/85–4/85
Daniel M. Burns	1/85–12/87
Joseph B. Keller	1/85–12/87
Audrey A. Terras	1/85–12/87
David A. Vogan, Jr.	1/85–12/87
Hyman Bass	1/86–12/86
Cora S. Sadosky[2]	1/86–12/87

[2]Bharucha-Reid died on 5 April 1985. Sadosky was elected to fill out the term.

James G. Arthur	1/86–12/88	Yiannis Moschovakis	1/87–
Jane P. Gilman	1/86–12/88	Linda A. Ness	1/87–
Chuu-Lain Terng	1/86–12/88	Marc A. Rieffel	1/87–
William A. Veech	1/86–	Carol S. Wood	1/87–
H. Blaine Lawson, Jr.	1/87–		

THE FUNCTIONS OF THE COUNCIL

The bylaws assign a number of powers and duties to the Council.
Article IV, Section 2 states that:

> The Council shall formulate and administer the scientific poli-
> cies of the Society and shall act in an advisory capacity to the
> Board of Trustees.

In fact, the administration lies with the secretary and the executive di-
rector. Much of the function of advising the Trustees is done through the
Executive Committee of the Council, which sits regularly with the Board of
Trustees at appropriate times.

Article IV, Section 8 begins by saying:

> The Council shall also have power to speak in the name of the
> Society with respect to matters affecting the status of mathematics
> or mathematicians, such as proposed or enacted federal or state
> legislation; conditions of employment in universities, colleges, or
> business, research or industrial organizations; regulations, policies,
> or acts of governmental agencies or instrumentalities; and other
> items which tend to affect the dignity and effective position of
> mathematics.

The existence of this bylaw is the result of a controversy discussed in
the chapter on political and social questions. The bylaw continues with a
description of operating procedures.

Other defined functions are the election of members and the setting of
times and places of meetings. These however have been delegated to the
Executive Committee or to the secretary and associate secretaries.

The nature of the work of the Council has changed with time. Some kinds
of business have moved to the Executive Committee. This has the advantage
that the Executive Committee and the Trustees meet jointly. Other business
has gravitated to the Providence office under the aegis of the executive director. The advantage here is that the Council does less administrative work
than in the past. The Council is more concerned with planning, as through its
Long Range Planning Committee, and with review, as through the periodic
review program of the Executive Committee, than with semi-administrative

matters such as the number of pages that a journal should publish. The objection of some Council members to this change in perspective, though they state it in other terms, is that it is harder to plan or to review effectively without the knowledge gained through immersion in operations.

The Council is governed by the Standard Code of Parliamentary Procedure (second edition) by Alice Sturgis. Here is the background. The Society had a Business Meeting in January 1974 in San Francisco at which it was anticipated that there might be parliamentary difficulties. Usually the Society had depended on the imperfect knowledge of parliamentary procedures of the president or the secretary. Occasionally a member was used as parliamentarian. In this instance it was decided that the seriousness of the situation warranted investigation of the possibility of using a professional parliamentarian. The name of Alice Sturgis was found in the San Francisco yellow pages. The Society could not have been more fortunate. Not only was Mrs. Sturgis the author of an excellent book based on sound, clearly stated principles but also she was experienced in advising groups. Moreoever, at the moment she professed to be interested in the special problems of non-profit organizations. With her guidance the meeting was handled effectively. The Council then adopted the book of Sturgis as the standard for the parliamentary governance of its meetings.

Presidents

The office of president has two aspects. First, it is a high honor that the Society bestows on one of its members. The president is an embodiment of mathematical virtue to be emulated and admired. The office has always been uncontested. Thus it is filled by the nominee of the Council, which is in turn confirming the selection of the nominating committee. One does not "campaign" for nomination. The writer has discovered no instance in which the name proposed was not instantly approved. This is not astonishing in that the pool of worthy candidates is large. There was a bit of muttering in the background when the first Jewish president was selected. At least one person who was the clear choice of the nominating committee declined to have his name presented to the Council.

In the year that E. J. McShane was the candidate, the Nominating Committee brought forward two names with the intent that the Council choose between them. The secretary at the time was John W. Green. He persuaded the committee that the proposed choice by the Council was a bad idea and the committee finally selected McShane. It was fourteen years before the name of the other candidate came to the Council for nomination.

The nature of the position of president as an honor is attested by the eminence of the holders as research mathematicians. There have been 49 presidents in the one hundred years of the Society. (One served for four years and each of the others for two.) Of these, 41 have been members of the National Academy of Sciences, including the last 21. Although it has never been a requirement, it is a qualification of which nominating committees have been aware.

Other than the duties that adhere to the title of president by common practice, there is one duty prescribed in the bylaws. That is to give a retiring presidential address within one year after completion of the term. However, the annual meeting was moved from December to January in 1958. The bylaw was not amended to allow for the resulting anomaly, but the address has most usually been given in the thirteenth month. It has been understood by some that the purpose of the bylaw is to assure that the office remain in the hands of persons who are primarily research mathematicians and capable of the required duty.

Second, the office is a working position. The burden has grown with time. It extends over the year as president-elect, two years as president, and a year as ex-president. The president is a member of the Executive Committee during all four years. During the second and third he is a member of the Board of Trustees and he sits with the Trustees in the first and fourth as well. These two bodies meet jointly for several days, usually in May and November, and occasionally at other times. The president is chairman of the Liaison Committee, which has continuing sporadic duties, including oversight of the work of the executive director. He is a member of the Agenda and Budget Committee, which has two planning sessions each year in preparation for the meetings of the Executive Committee and Board of Trustees. The president is a member of the Committee on Committees, which advises on presidential appointments as a source of ideas. He represents the Society in the Council of Scientific Society Presidents, a "club" of such individuals from many disciplines, and on the Conference Board of the Mathematical Sciences, an umbrella organization.

The president is the representative of the Society to the outside world in matters of moment. Although a great deal of the business of the Society is conducted by the executive director, the treasurer, or the secretary, in matters where the prestige of the Society or the dignity of the profession is in question, it is the president who speaks for the Society.

There are clearly more people who are worthy of the office, would grace it, and would perform effectively than the Society can use at the rate of one every two years. The proposal has been advanced casually that a partial remedy is to reduce the term to one year. The counter argument has prevailed that were this the case the entire term would be spent learning the job.

Some presidents have not had much continuing experience with the business of the Society, having perhaps served on the Council as a journal editor, been a member of a few committees, and been an elected vice-president, none of this in the recent past. On the other hand the wisdom and sophistication of presidents in the context of the general mathematical community is substantial.

In the days before the prevalence of airplane travel, the geographic location of the president could present problems. Presidents came from no further west nor less convenient location than Chicago up until E. R. Hedrick from UCLA in 1929–1930, R. L. Moore from the University of Texas in 1937–1938, and G. C. Evans from the University of California at Berkeley in 1940. The Trustees voted to pay for one round trip transcontinental rail fare per year for Evans.

Until 1948 the president remained on the Council for six years beyond the term of office, so that the number of such Council members was usually three. Then there was a change in the bylaws setting that term of continued service at one year. The president is eligible for other office, say editor or member-at-large after his term is over. In fact, since the change in bylaws, it appears that no president has served as member-at-large or journal editor.

In the following pages there is a picture of each president and a brief curriculum vitae.

GRIFFITH CONRAD EVANS
PRESIDENT 1939–1940

G. C. Evans was born in Boston on 11 May 1887. He took his A.B. in 1907, his M.A. in 1908, and his Ph.D. in 1910, all at Harvard. His thesis adviser was W. F. Osgood. His principal academic appointments were at Rice from 1912 to 1934 and at the University of California, Berkeley from 1934 to 1954. He gave Colloquium Lectures in 1916. He was a member of the American Academy of Arts and Science and of the National Academy of Sciences. He died on 8 December 1973.

HAROLD CALVIN MARSTON MORSE
PRESIDENT 1941–1942

Marston Morse was born on 24 March 1892 in Waterville, ME. His B.A. was from Colby College in 1914. He received the M.A. in 1915 and the Ph.D. in 1917, both from Harvard. His thesis adviser was G. D. Birkhoff. His principal academic appointments were at Harvard from 1926 to 1935 and at the Institute for Advanced Study from 1935 until retirement in 1975. He gave the Colloquium Lectures of 1931 and the Gibbs Lecture of 1952. He received the Bôcher Prize in 1933. He was chairman of the Division of Mathematics of the National Research Council in 1950–1952, a member of the National Science Board in 1950–1954, and chairman of the U.S. National Commission for Mathematics in 1959–1963. He received the President's Certificate of Merit in 1947, the National Medal of Science in 1965, and the Croix de Guerre and was a Chevalier of the Légion d'Honneur. He was a member of the American Academy of Arts and Sciences, a corresponding member of the Italian National Academy of Lincei, an associate member of the French Academy of Sciences, and a member of the National Academy of Sciences. He died on 22 June 1977.

Marshall H. Stone

Marshall Harvey Stone
President 1943–1944

Marshall Stone was born in New York, NY on 8 April 1903. He received his A.B. from Harvard in 1922 and his Ph.D. in 1926. His thesis adviser was G. D. Birkhoff. His principal academic appointments have been at Harvard, 1927–1931 and 1933–1946, at Chicago from 1947 to 1968, and at the University of Massachusetts, Amherst from 1968. He gave Colloquium Lectures in 1939 and the Gibbs Lecture of 1956. He held many visiting academic appointments and served on many international commissions. He was president of the International Mathematical Union in 1952–1954 and president of the International Committee on Mathematics Instruction, 1959–1962. He is a member of the National Academy of Sciences.

Theophil Henry Hildebrandt
President 1945–1946

T. H. Hildebrandt was born in Dover, OH on 24 July 1888. He received his A.B. from the University of Illinois in 1905, his S.M. from the University of Chicago in 1906 and his Ph.D. from the same institution in 1910 under E. H. Moore. He spent the major portion of his professional career at the University of Michigan, where he went as an instructor of mathematics in 1909. He served as chairman of the department from 1934 until his retirement in 1957. Professor Hildebrandt received the Chauvenet Prize of the Mathematical Association of America in 1929. He died on 9 October 1980.

Einar Hille
President 1947–1948

Einar Hille was born on 28 June 1894 in Stockholm. He received the A.B. in 1913 from the University of Stockholm, the M.Ph. in 1914, the Lic. Ph. in 1916, and the Ph.D. in 1918. His principal academic appointments were at Princeton from 1922 to 1933 and at Yale from 1933 until his retirement in 1962. He gave the Colloquium Lectures of 1944. His dissertation won a Mittag-Leffler Prize. He was a member of the American Academy of Arts and Sciences, the Swedish Academy, and the National Academy of Sciences. He died on 12 February 1980.

JOSEPH LEONARD WALSH
PRESIDENT 1949–1950

Joseph L. Walsh was born on 21 September 1895 in Washington, D.C. He received the degrees of S.B. in 1916 and the Ph.D. in 1920 from Harvard. His thesis adviser was Maxime Bôcher. His principal academic appointment was at Harvard from 1921 to his retirement in 1966. He then took an appointment at the University of Maryland. He was chairman of the organizing committee of the International Congress of Mathematicians in 1950. He was a member of the National Academy of Sciences. He died on 10 December 1973.

John von Neumann

John von Neumann
President 1951–1952

John von Neumann was born in Budapest on 28 December 1903. He received an undergraduate degree in chemical engineering from the Eidgenössische Techniche Hochschule and a Ph.D. in mathematics from the University of Budapest both in 1926. His principal academic appointment was as professor at the Institute for Advanced Study from 1933 until his death on 8 February 1957. He was a member of the Atomic Energy Commission. He received the Bôcher Prize in 1937, the Medal for Merit in 1947, and the Medal of Freedom, the Albert Einstein Commemorative Award, and the Eurico Fermi Award all in 1956. He was a member of the National Academy of Sciences.

Gordon Thomas Whyburn
President 1953–1954

G. T. Whyburn was born on 7 January 1904 in Lewisville, TX. He received the A.B. in 1925, the M.A. in chemistry in 1926, and the Ph.D. in mathematics in 1927, all from the University of Texas. His thesis adviser was R. L. Moore. His principal academic appointment was at the University of Virginia, where he went as chairman in 1934, serving until 1966, and where he remained until his death on 8 September 1969. He was the Colloquium Lecturer of 1940. He received the Chauvenet Prize of the Mathematical Association of America. He was a member of the National Academy of Sciences.

Raymond Louis Wilder
President 1955–1956

Raymond L. Wilder was born on 3 November 1896 in Palmer, MA. He received a B.Phil. from Brown in 1920 and an M.Sc. in actuarial mathematics in 1921. He received his Ph.D. at the University of Texas. His thesis adviser was R. L. Moore. His principal academic appointment was at the University of Michigan where he served from 1926 to 1967 and where he was research professor from 1947. Following this he became a research associate at the University of California, Santa Barbara. He gave the Colloquium Lectures of 1943 and the Gibbs Lecture of 1969. He was president of the Mathematical Association of America in 1965–1966. He received the Lester R. Ford Award and the Distinguished Service Award of the MAA in 1973. He was a member of the National Academy of Sciences. He died on 7 July 1982.

Richard Brauer

Richard Dagobert Brauer
President 1957–1958

Richard D. Brauer was born on 10 February 1901 in Berlin. He received the Ph.D. from the University of Berlin in 1925. His principal academic appointments were at the University of Toronto from 1935 to 1948, at the University of Michigan from 1948 to 1951, and at Harvard from 1951 until 1971. He received the Cole Prize in algebra in 1949. He gave the Colloquium Lectures of 1948. He received the National Medal for Scientific Merit in 1970. He was a member of the National Academy of Sciences. He died on 17 April 1977.

E. J. McShane

Edward James McShane
President 1959–1960

E. J. McShane was born on 10 May 1904 in New Orleans. He received B.E. and B.S. degrees from Tulane in 1925 and the M.S. in 1927. He received the Ph.D. from the University of Chicago in 1930. His thesis adviser was L. M. Graves. His principal academic position was at the University of Virginia from 1935 to 1974. He gave the Colloquium Lectures of 1943. He was president of the Mathematical Association of America in 1953–1954. He is a member of the National Academy of Sciences.

Deane Montgomery
President 1961–1962

Deane Montgomery was born on 2 September 1909 in Weaver, MN. He received his B.A. from Hamline in 1929, his M.S. from the University of Iowa in 1930 and then his Ph.D. in 1933. His thesis adviser was E. W. Chittenden. His principal appointments were at Smith College from 1935 to 1946 and at the Institute for Advanced Study, where he became permanent member in 1948 and professor in 1951 until his retirement in 1980. He gave the Colloquium Lectures of 1951. He was president of the International Mathematical Union in 1974–1978. He is a member of the National Academy of Sciences.

Joseph Leo Doob
President 1963–1964

Joseph L. Doob was born in Cincinnati on 27 February 1910. He received his A.B. in 1930, his A.M. in 1931, and his Ph.D. in 1932, all from Harvard. His thesis adviser was Joseph L. Walsh. His principal academic appointment was at the University of Illinois from 1935 until his retirement in 1977. He gave the Colloquium Lectures of 1959. He was president of the Institute of Mathematical Statistics in 1950. He received the National Medal of Science in 1980. He is a foreign associate of the Academy of Sciences, Paris and a member of the American Academy of Arts and Science and the National Academy of Sciences.

A. Adrian Albert

Abraham Adrian Albert
President 1965–1966

A. A. Albert was born on 9 November 1905 in Chicago, IL. He received the B.S. in 1926, the M.S. in 1927, and the Ph.D. in 1928, all from the University of Chicago. His thesis adviser was Leonard Eugene Dickson. His principal academic appointment was at the University of Chicago, where he served from 1931 until his death on 6 June 1972. He was dean of the Division of Physical Sciences from 1962 to 1971. He was awarded a Cole Prize in algebra in 1939 and gave a set of Colloquium Lectures in 1939 as well. He was chairman of the Committee to Prepare a Budget for Mathematics for the National Science Foundation in 1950, chairman of the Committee on a Survey of Training and Research Potential in the Mathematical Sciences (the Albert Survey), 1955–1957. He was associated with the Institute for Defense Analyses as trustee and as director for a year. He was chairman of the Consultative Committee of the International Congress in Nice and vice-president of the International Mathematical Union. He was a member of the National Academy of Sciences. He died on 6 June 1972.

Charles B. Morrey, Jr.

Charles Bradford Morrey, Jr.
President 1967–1968

C. B. Morrey was born in Columbus on 23 July 1907. He received his A.B. from Ohio State in 1927. His M.A. in 1928 and his Ph.D. in 1931 were both from Harvard. His thesis adviser was G. D. Birkhoff. His principal academic appointment was at the University of California, Berkeley from 1933 until his retirement in 1977. He gave the Colloquium Lectures of 1964. He was a member of the National Academy of Sciences. He died on 29 April 1984.

Oscar Zariski

Oscar Zariski
President 1969–1970

Oscar Zariski was born on 24 April 1899 in Kobrin, USSR. He received the degree of Doctor of Mathematics from the University of Rome in 1924. His thesis adviser was Guido Castel nuovo. His principal academic appointments were at Johns Hopkins, 1927–1945 and at Harvard from 1947 until his retirement in 1969. He gave the Colloquium Lectures of 1947. He received the Cole Prize in 1944 and the Steele Prize in 1981. He was awarded the National Medal of Science in 1965 and the Wolf Prize in 1982. He was a member of the National Academy of Sciences. He died on 4 July 1986.

Nathan Jacobson
President 1971–1972

Nathan Jacobson was born in Warsaw on 8 September 1910. He received his A.B. from the University of Alabama in 1930 and his Ph.D. from Princeton in 1934. His thesis adviser was J. H. M. Wedderburn. His principal academic appointment has been at Yale from 1947 until his retirement in 1981. He gave the Colloquium Lectures of 1955. He is a member of the American Academy of Arts and Sciences and of the National Academy of Sciences.

Saunders Mac Lane
President 1973–1974

Saunders Mac Lane was born on 4 August 1909 in Norwich, CT. He received a Ph.B. from Yale in 1930, an M.A. in mathematics from Chicago in 1931, and a D. Phil. from the Mathematisches Institut, Göttingen in 1934. His thesis was written under the direction of Paul Bernays. His principal academic appointments were at Harvard from 1938 to 1947 and at Chicago from 1947 until his retirement in 1982. He gave the Colloquium Lectures of 1963. He was president of the Mathematical Association of America in 1951–1953; vice-president of the American Philosophical Society, 1973–1974; vice-president of the National Academy of Sciences, 1973–1981; and member of the National Science Board, 1974–1980. He received the Chauvenet Prize in 1941 and the Distinguished Service Award in 1975, both from the Mathematical Association of America, and the Steele Prize in 1986.

Lipman Bers

Lipman Bers
President 1975–1976

Lipman Bers was born in Riga on 22 May 1914. He was awarded the degree of Doctor of Natural Sciences at the University of Prague in 1938. His thesis adviser was Charles Loewner. His principal academic appointments were at Syracuse, 1945–1951, New York University, 1951–1964, and Columbia from 1964 until his retirement in 1982. He gave a set of Colloquium Lectures in 1971. He received the Steele Prize in 1975. He is a member of the American Academy of Arts and Sciences and of the National Academy of Sciences.

R H Bing
President 1977–1978

R H Bing was born on 20 October 1914 in Oakwood, TX. He took his B.S. from Southwest State Teachers College in 1935. His M.Ed. in 1938 and his Ph.D. in 1945 were both from the University of Texas. His thesis adviser was R. L. Moore. His principal academic appointments were at the University of Wisconsin from 1943 to 1973 and at the University of Texas from 1973. He gave the Colloquium Lectures of 1970. He was a member of the National Science Board in 1968–1975. He was a member of the National Academy of Sciences. He died on 27 April 1986.

Peter David Lax
President 1979–1980

Peter Lax was born on 1 May 1926 in Hungary. He received his A.B. in 1947 and his Ph.D. in 1949 from New York University. His thesis adviser was K. O. Friedrichs. His principal academic appointment has been at the Courant Institute beginning in 1949. He gave the von Neumann Lecture for the Society for Industrial and Applied Mathematics in 1968 and the Hedrick Lectures for the Mathematical Association of America in 1973. He gave the Colloquium Lectures in 1987. He received the Wiener Prize in 1975. He was a member of the National Science Board in 1980–1986. He is a member of the National Academy of Sciences.

Andrew M. Gleason

Andrew Mattel Gleason
President 1981–1982

Andrew Gleason was born on 4 November 1921 in Fresno, California. He received the B.S. degree from Yale in 1942. He was a Junior Prize Fellow at Harvard. His principal academic appointment was at Harvard since 1950. He is a member of the American Academy of Arts and Sciences and of the National Academy of Sciences.

Julia B. Robinson

Julia Bowman Robinson
President 1983–1984

Julia Robinson was born on 8 December 1919 in St. Louis, MO. She received the A.B. in 1940, the M.A. in 1941, and the Ph.D. in 1948, all from the University of California, Berkeley. Her thesis adviser was Alfred Tarski. Her major appointment, received belatedly because of nepotism rules, was as professor at the University of California, Berkeley. She held a MacArthur Prize Fellowship. She gave Colloquium Lectures in 1980. She was a member of the National Academy of Sciences. She died on 30 July 1985.

Irving Kaplansky

Irving Kaplansky
President 1985–1986

Irving Kaplansky was born on 22 March 1917 in Toronto, Ontario. He received a B.A. in 1938 and an M.A. in 1939, both from Toronto, and a Ph.D. from Harvard in 1941. His thesis was written under the direction of Saunders Mac Lane. He served on the faculty of the University of Chicago from 1945 until 1984, when he became the second director of the Mathematical Sciences Research Institute in Berkeley. He was the highest ranking candidate on the first Putnam examination. He is a member of the National Academy of Sciences.

G. D. Mostow

GEORGE DANIEL MOSTOW
PRESIDENT 1987–1988

G. D. Mostow was born on 4 July 1923 in Boston. He received the B.A. in 1943, the M.A. in 1946, and the Ph.D. in 1948, all from Harvard. His thesis adviser was Garrett Birkhoff. His principal academic appointments have been at Johns Hopkins from 1952 to 1961 and at Yale since 1961. He gave Colloquium Lectures in 1979. He served as chairman of the U.S. National Committee for Mathematics in 1972–1973, as chairman of the Office of Mathematical Sciences for the National Academy of Sciences-National Research Council, as a member of the Executive Commitee of the International Mathematical Union, 1983–1987, and on the Board of Trustees of the Institute for Advanced Study. He is a member of the National Academy of Sciences.

Officers

There is a chapter devoted to the presidents of the second fifty years. The other officers who are Council members are the vice-presidents, members of the secretariat, the treasurer and associate treasurer, and the librarian while the position existed. These are discussed here.

Vice-President

Vice-presidents have been chosen in a contested election since 1941. The office of vice-president is not well defined by function. The bylaws provide a rule of succession in that should the president die or resign when there is a president-elect then that person shall succeed at once. Only in the years that there is no president-elect does the Council, with the approval of the Trustees, make a selection among the vice-presidents. Once in the last fifty years a situation arose where the spirit of these rules had to be considered. During the illness of Julia Robinson toward the end of her term, President-Elect Irving Kaplansky was designated acting president. Vice-presidents used to be called upon frequently to fill in for absent presidents. This phenomenon corresponds to complications of WWII and to difficulties in transportation before plane travel was well established. Since 1962 such instances have been rare. For several months in 1943 during the absence of President M.H. Stone from the country, Vice-president C. C. MacDuffee was named acting president by the Council. There have been occasions when the Council chose a presiding officer other than a vice-president. In September 1952, neither the president nor any vice-president was present and Ex-president T. H. Hildebrandt was chosen to preside. Ex-president Richard Brauer presided in August 1956 even though a vice-president was present and he presided again in December 1956. In August 1960 neither the president nor any vice-president was present and the Council elected E. G. Begle to preside. At the Summer meeting of 1969 in Eugene in the absence of President Zariski the Council elected Ex-president C. B. Morrey to the chair, though a vice-president was present, and he presided at the Business Meeting as well. At the Summer Meeting of 1970 in Laramie neither the president nor any vice-president was present.

The secretary convened the Council, which in turn elected J. T. Schwartz to preside. He then presided also at the Business Meeting.

Vice-presidents are occasionally assigned duties by the president, typically to preside over panel discussions on matters of Society policy. There is no such office in the bylaws as vice-president for an area of Society activity.

A vice-president is not eligible for immediate re-election to that office. In fact, no person has held the office twice. There has been some sentiment against it. In August 1954 (and there may be similar instances) the Nominating Committee proposed the name of E. J. McShane for vice-president. It was pointed out that he had already served as vice-president and the chairman of the nominating committee, C. B. Morrey, Jr., withdrew the name in favor of H. F. Bohnenblust. In fact Bohnenblust was not elected until a later time. Few persons were elected president without having served as vice-president. Since the first fourteen years of the Society the only ones are M. H. Stone, R. Brauer, and C. B. Morrey.

Here are the Vice-Presidents since 1938:

R. E. Langer	1/38–12/39	Emil Artin	1/55–12/56
J. F. Ritt	1/38–12/39	A. A. Albert	1/56–12/57
C. R. Adams	1/39–12/40	Nathan Jacobson	1/56–12/57
T. C. Fry	1/40–12/41	Salomon Bochner	1/57–12/58
F. D. Murnaghan	1/40–12/41	Garrett Birkhoff	1/58–12/59
T. Y. Thomas	1/41–12/42	Norman E. Steenrod	1/58–12/59
C. C. MacDuffee	1/42–12/43	Ralph P. Boas	1/59–12/60
J. D. Tamarkin	1/42–12/43	H. F. Bohnenblust	1/60–12/61
L. M. Graves	1/43–12/44	Oscar Zariski	1/60–12/61
Einar Hille	1/44–12/45	M. R. Hestenes	1/61–12/62
G. T. Whyburn	1/44–12/45	William Feller	1/62–12/63
J. M. Thomas	1/45–12/46	Andrew M. Gleason	1/62–12/63
L. R. Ford	1/46–12/47	S. S. Chern	1/63–12/64
Saunders Mac Lane	1/46–12/47	Lipman Bers	1/63–12/64
P. A. Smith	1/47–12/48	George W. Mackey	1/64–12/65
John von Neumann	1/48–12/49	Mark Kac	1/65–12/66
Hassler Whitney	1/48–12/49	Samuel Eilenberg	1/66–12/67
W. T. Martin	1/40–12/50	Norman Levinson	1/66–12/67
E. J. McShane	1/50–12/51	R H Bing	1/67–12/68
R. L. Wilder	1/50–12/51	David Blackwell	1/68–12/69
G. A. Hedlund	1/51–12/52	Kurt O. Friedrichs	1/68–12/69
Richard Brauer	1/52–12/53	H. S. MacDonald Coxeter	
Deane Montgomery	1/52–12/53		1/69–12/70
D. H. Lehmer	1/53–12/54	Peter D. Lax	1/70–12/71
L. V. Ahlfors	1/54–12/55	Isadore M. Singer	1/70–12/71
Antoni Zygmund	1/54–12/55	Edwin H. Spanier	1/71–12/72

Richard D. Anderson	1/72–12/73	Mary Ellen Rudin	1/80–12/81
R. Creighton Buck	1/72–12/73	Paul R. Halmos	1/81–12/82
Edwin E. Moise	1/73–12/74	Michael Artin	1/82–12/83
Raoul H. Bott	1/74–12/75	Elias M. Stein	1/82–12/83
Irving Kaplansky	1/74–12/75	Calvin C. Moore	1/83–12/84
John W. Milnor	1/75–12/76	Jacob T. Schwartz	1/84–12/85
Mary W. Gray	1/76–12/77	Stephen Smale	1/84–12/85
Louis Nirenberg	1/76–12/77	Linda Preiss Rothschild	1/85–12/86
William Browder	1/77–12/78	Richard A. Askey	1/86–12/87
Julia B. Robinson	1/78–12/79	Olga Taussky-Todd	1/86–12/87
George W. Whitehead	1/78–12/79	Karen Uhlenbeck	1/87–
George Daniel Mostow	1/79–12/80	Barry Simon	1/88–
Hyman Bass	1/80–12/81	William P. Thurston	1/88–

SECRETARIES

The bylaws specify that there shall be a secretary, who is a member of the Council, but prescribe almost no duties. The secretary or an associate secretary notifies Council members of the time and place of the Annual Meeting. The secretary is one of those who can call a meeting of the Trustees. The secretary is a member of the Executive Committee and of the Liaison Committee, whose function is the "immediate direction" of the executive director. The secretary by inference conducts elections in that the secretary is responsible for the sending of ballots. The secretary is responsible for the election of members to the extent of supplying an application blank.

The Council has made up for this deficiency in definition by both general statements and specific assignments. A Committee on the Reorganization of the Secretariat, consisting of L. M. Graves, chairman, W. B. Fite, J. R. Kline, C. C. MacDuffie, E. J. McShane, W. T. Martin, and C. B. Morrey made its final report to the Council of 30 December 1941. In the course of declining to recommend a delimitation or restructuring of the office of secretary, it made the following observations:

> It is desirable for the secretary to maintain some general oversight over all the activities of the Society. His is the responsibility of interpreting the policies of the Council and putting these into action, and for interpreting to the Council the reasons for the actions of the Council itself in the past so as to keep a certain degree of continuity of policy, or at least to guarantee that discontinuities occur only after careful consideration.

> It is desirable that some general supervision of the financial affairs of the Society remain in the hands of the secretary.

As the Society has grown in size and its activities have increased in complexity, the work of the Society has become more specialized and compartmentalized but these principles seem still to stand.

The secretary from 1921 to 1940 was R. G. D. Richardson, who was dean of the graduate school at Brown University since 1926. See [A] for a picture and an account of his life. Dean Richardson retired in 1942 and died in 1949.

There have been instances of temporary secretaries, including Arnold Dresden for J. R. Kline during part of 1943 while Kline was ill and Lowell Paige for J. W. Green during part of 1961 while Green was on leave. Orville Harrold replaced Everett Pitcher during part of a meeting in January 1976 when the Council discussed a matter concerning Lehigh University.

John Robert Kline

John Robert Kline was secretary from 1941 to 1950. He took his Ph.D. in 1916 at the University of Pennsylvania under R. L. Moore and was on the faculty there from 1920. He served at various times as an associate editor of the *Transactions*, the *Bulletin*, and the *American Journal* and on the Editorial Board of the Colloquium Publications. He was associate secretary in 1933–1936 and secretary of the International Congress of 1950. His death came in 1955.

Edward Griffith Begle

Edward Griffith Begle was secretary in 1951–1956. He was born in 1914 and took his Ph.D. under Solomon Lefschetz at Princeton in 1940. He was on the mathematics faculty at Yale from 1942 to 1961 and at Stanford as professor and director of the School Mathematics Study Group until his death in 1978.

E. G. Begle

JOHN WILLIE GREEN

John Willie Green was secretary in 1957–1966. He was born in 1914 and took his Ph.D. at the University of California, Berkeley in 1938 under G. C. Evans. He was on the faculty of the University of California, Los Angeles from 1945 until his retirement in 1984. He was associate secretary in 1947–1955 and was the principal investigator for the Albert survey in 1956–1957.

Arthur Everett Pitcher

Arthur Everett Pitcher was secretary from 1967 to 1988. He was born in 1912 and took his Ph.D. in 1935 at Harvard under Marston Morse. He was on the faculty of Lehigh University from 1938 until his retirement in 1978. He served as associate secretary from 1959 to 1966.

Associate Secretaries

The associate secretaries are in charge of programs of the Society. The bylaws state that "papers intended for presentation at any meeting of the Society shall be passed upon in advance by a program committee appointed by or under the authority of the Council; and only such papers shall be presented as shall have been approved by such committee." Formally, the committee with respect to Annual and Summer Meetings is the Program Committee and with respect to regional meetings is the Committee to Select Hour Speakers for the region. In practice, the associate secretary of the region or assigned to a national meeting alone accepts or occasionally rejects ten-minute contributed papers and defers to the organizer of a special session on papers for that session. The associate secretary is free to consult referees and refers the occasional paper about which he is in doubt to the relevant committee.

For many years, the associate secretary in the area where a national meeting fell was assigned to the meeting. It developed that the load was unevenly distributed, to the point that one might be starting work on one national meeting before finishing work on another. A rotation was established among associate secretaries beginning in 1979. The effect is that an associate secretary handles one national meeting in two years, and that not until the second year of a term, so that the associate secretary has gathered experience and is known well in advance.

For many years, through 1960, papers presented by title were part of the program of a meeting in a section called Supplementary Program and the abstracts were passed upon by the cognizant associate secretary. This system was changed so that the papers were no longer associated with a meeting and the associate secretaries handled them in rotation in batches of about fifty.

In 1938 there were three regions, East, West, and Far West, with an associate secretary for each. As noted in the section on the rise of the office of executive director, there was also an associate secretary for financial affairs. This had been done informally but was a formal office from 1944 to 1949.

Geographical regions were not as well-defined then as now. Thus one finds T. R. Hollcroft, whose home base was Wells College, handling consecutive programs in New York and Chicago and A. C. Schaeffer, based at Stanford, with a program in Chicago. The designation West was replaced by Central at the behest and during the term of P. T. Bateman. The Southeast was not a designated region for meetings until 1948 and the first meeting there was the Annual Meeting on 27–29 December 1950 in Gainesville.

Here are the associate secretaries of the second fifty years:

East

T. R. Hollcroft	Jan 1937–Dec 1950
L. W. Cohen	Jan 1951–Dec 1954
R. D. Schafer	Jan 1955–Dec 1958
Everett Pitcher	Jan 1959–Dec 1966
Herbert Federer	Jan 1967–Dec 1968
Leonard Gillman	Jan 1969–Dec 1970
Walter Gottschalk	Jan 1971–Dec 1976
Raymond G. Ayoub	Jan 1977–Dec 1982
W. Wistar Comfort	Jan 1983–

Far West

T. M. Putnam	Jan 1930–Dec 1942
A. D. Michal[1]	Jan 1943–Dec 1944
A. C. Schaeffer	Jan 1945–July 1947
John W. Green	July 1947–Apr 1955
Victor L. Klee, Jr.[2]	May 1955–Dec 1959
Richard S. Pierce	Jan 1960–Dec 1970
Kenneth A. Ross	Jan 1971–Dec 1981
Hugo Rossi	Jan 1982–Dec 1987
Lance W. Small	Jan 1988–

West

W. L. Ayres	Jan 1938–Dec 1943
G. B. Price[3]	Jan 1944–Dec 1944
R. H. Bruck[4]	Jan 1945–Sept 1948
J.W.T.Youngs[4]	Oct 1948–Dec 1963
Seymour Sherman	Jan 1964–Dec 1966
Paul T. Bateman	Jan 1967–Dec 1983
Robert W. Fossum	Jan 1984–Dec 1987
Andy Roy Magid	Jan 1988–

---

[1]Resigned before completing his term.

[2]Pierce acted for Klee during part of 1958.

[3]Although he was elected, he was unable to serve because of civilian duties with the military forces in England. J.W.T. Youngs served as acting associate secretary briefly.

[4]Bruck resigned effective with the Summer Meeting of 1948 and Youngs filled out his term.

Southeast

W. M. Whyburn	Jan 1950–Dec 1953
J. H. Roberts	Jan 1954–Dec 1957
G. B. Huff	Jan 1958–Dec 1963
Morton L. Curtis	Jan 1964–Dec 1964
Orville G. Harrold, Jr.	Jan 1965–Dec 1976
Frank T. Birtel	Jan 1977–

Associate secretary for Financial Affairs

M. H. Ingraham	Jan 1938–Dec 1942
W. L. Ayres	Jan 1943–Sept 1946
G. B. Price	Sept 1946–Dec 1949

Treasurers and Associate Treasurers

The position of treasurer has changed materially with time. It was once the case that the treasurer kept the books of the Society or closely supervised an employee who did so. With the advent of the executive director, this aspect of supervision passed to the central office. The treasurer remained responsible for fiscal policy though increasing assistance became available with the employment of higher grades of professionals in the fiscal area.

Beginning in 1938, the sequence of treasurers has been as follows:

B. P. Gill	1938–1948
A. E. Meder, Jr.	1949–1964
W. T. Martin	1965–1973
F. P. Peterson	1973–

In June 1965, the Trustees created the position of assistant treasurer, elected by the Trustees among their own membership. The assistant treasurer at the request of the Liaison Committee could act as an alternate for the treasurer. In December 1965 E. G. Begle was elected assistant treasurer for 1967 and also Trustee for the unexpired term of C. B. Morrey, who became president. M. H. Protter was assistant treasurer in 1968–1972. In August 1972 the Business Meeting ratified an amendment to the bylaws creating the post of associate treasurer and enlarging the Board and the Council so that the position carried membership on each. Protter held the elected position through 1976 and Steve Armentrout has held it since 1977.

It was intended that either the treasurer or the associate treasurer be able to handle the duties of the office and that one be available at all times. In addition there has been by agreement a division of labor in that the associate treasurer has supervised the budget process at *Mathematical Reviews* and the setting by the management of salaries and wages.

THE LIBRARIAN

Until the time of the sale of the library, the librarian was a member of the Council. The position was filled by R. C. Archibald from 1921 to 1941, by Arnold Dresden from 1942 to 1950, and by Jekuthiel Ginsburg in 1951 and 1952. See the section on the change in location of the office of the Society from New York to Providence, where the sale of the library and the consequent elimination of the post of librarian are discussed.

THE PROFESSIONAL DIRECTORY

For many years the list of officers and of committees was published with the bylaws in the *Bulletin*. In 1954 an *Administrative Directory* appeared, showing not only officers and committees but staff and other information. It also carried the officers of the National Research Council Division of Mathematics and the Mathematical Association of America. In 1958 the officers of the Mathematics Conference Organization (later the Conference Board) appeared and in 1960 those of the Association for Symbolic Logic. The year 1960 is the last one in which the bylaws and articles of incorporation appeared in that publication.

The directory was published annually and expanded rapidly to include more organizations and more information. In 1983 the name was changed to *Professional Directory*. By 1987 it listed officers and committees of thirty-seven organizations related to the mathematical sciences. It identified key individuals in seven government agencies. It identified editors and publishers of mathematics journals. It gave mailing addresses and telephone numbers of hundreds of named individuals. It identified about 1800 educational institutions and indexed them. The publication is a privilege of institutional membership and is distributed to all persons listed in it.

Executive Directors

The burden of work on the secretary and the creation of the post of executive director are closely related. In 1938, the office of the Society at Columbia University consisted of four persons, of whom one was office manager, Miss Evelyn Hull. The secretary, since 1921, was R. G. D. Richardson, who had signified his intent to retire from that position and in fact did formally tender his resignation at the Council of 27 December 1939. In recognition of the increasing load on the secretary, a committee with L. M. Graves as chairman recommended in September 1940 inter alia that the secretary, when and if he deemed it desirable, should request the Council that provision be made for another associate secretary or assistant secretary at the same or a neighboring institution. This was not the first time that such relief had been instituted, for Mark H. Ingraham was already associate secretary with a variety of assigned duties quite different from the supervision of scientific programs, such as fund raising. The committee however recommended in favor of flexibility of assignment of duties among the secretary, associate secretaries, and the treasurer and against the formalization of a position of associate secretary for financial affairs.

Notwithstanding this recommendation the Council of 26 November 1943 authorized the designation of one of the associate secretaries as associate secretary for Financial Affairs.

As the country emerged from World War II it was apparent that more extensive reorganization of the Society was required. The Council of 16 April 1948 received a report from a Committee on Reorganization, with W. L. Ayres as chairman. Among the changes approved by the Council was one to divide the office of secretary into two offices, executive director and secretary. The former was to be a full-time paid employee of the Society, "in charge of the central office of the Society and charged with the general administration of the affairs of the Society once policies have been set by the Council and its Executive Committee." Duties formerly performed by the associate secretary for Financial Affairs devolved on the executive director and the former office was eliminated. The report specified that the executive director was to be nominated by the Council and approved by the Trustees and the bylaws concerning the executive director initially so stated. It was belatedly noted that the Trustees control Society funds and should be the ones to execute an employment contract. A change in bylaws in 1960 specified that the executive director is appointed by the Trustees with approval of the Council. It was deemed desirable that the executive director be a mathematician.

In creating the post of executive director and in dividing the duties, that committee made the following statement:

> The secretary of the Society will continue to be the principal officer of the Society concerned with policy making. He will work with the president, the Council, and its committees in setting the policies of the Society. While the secretary and the executive director must cooperate closely in their work, the division of their functions can be described briefly in the two phrases "policy making" and "administration." Once the Council has set the policy of the Society it will be the duty of the executive director to carry out the administration.

It was the stated intent of the committee, approved by the Council, to relieve the pressure on the secretary and to make the job manageable.

A committee consisting of J. R. Kline, chairman, W. L. Ayres, and W. T. Martin was appointed by President Einar Hille to nominate a candidate for executive director. The committee gave serious consideration to about a dozen names. At the Council of 30 August 1949, the committee offered the name of Holbrook M. MacNeille, whom the Council approved subject to approval by the Board of Trustees, which had already been secured in essence. MacNeille began work on 14 November 1949. Toward the end of a five-year term he was commended and recommended for reappointment but signified his wish not to be reappointed.

Along with the recommendation for reappointment came the recommendation, subsequently incorporated in the bylaws, that a Liaison Committee, consisting of the president, the secretary and the treasurer, be appointed to advise the executive director between meetings of the Council or the Board of Trustees. Moreover, in background correspondence and incorporated in the letter of appointment to the second executive director, it is indicated that communication with the Liaison Committee, except in purely financial matters, should be through the secretary.

The next executive director, beginning on 1 September 1954, was John H. Curtiss. In the course of the search, there were at least twenty-five names considered, several quite seriously. Curtiss was elected to a two-year term followed by a three-year term. Although he was offered reappointment he declined.

The third executive director was Gordon L. Walker, elected in 1959. At this time there was an advertised search. He was offered a two-year term, which was followed by continuing appointments until his retirement in the middle of 1977.

The fourth executive director was William J. LeVeque who served from 1 July 1977 until his retirement toward the end of 1988. His appointment again followed a substantial advertised search.

The post of executive director is difficult. One wants a competent mathematician of sufficient stature to be respected by mathematicians but the duties require empathy with mathematicians and administrative skills rather than substantive mathematics. Some executive directors and associates have tried to arrange time to continue mathematical work and the Society Trustees and Brown University have been receptive to adjunct appointments to facilitate this but it appears never to have worked.

The increasing scope and bulk of the position of executive director can be seen from the following summary figures. The chosen years correspond to the changes in executive director, except the last.

Year	Number of Members	Number of Employees	Income
1948	3805	10	$ 145,000
1954	4680	20	294,000
1959	6652	63[1]	509,400
1977	16987	177	5,187,000
1987	20504	218	13,430,000

These figures are presented quite diversely in the records because fiscal years and reporting formats change with time, so that strictly comparable figures are not available. In addition to the change in number of members, which is apparent, there is a large increase in activity in publication that affects both number of employees and income and there is substantial inflation.

The position of executive director grew in complexity to the point that assistants with varying titles were appointed. These were

Lincoln K. Durst, Deputy Director
1970–1984

Jill P. Mesirov,[2] Assistant and then
Associate Executive Director
1982–1985

James W. Maxwell[3] Associate Executive Director
1984–

James A. Voytuk, Associate Executive Director
1985–

---

[1] 1960–61 budget

H. M. MacNeille

Holbrook Mann MacNeille

Holbrook Mann MacNeille was born in New York on 11 May 1907. He graduated with highest honors from Swarthmore College in 1928. He was a Rhodes Scholar at Balliol College, Oxford in 1928–1930 with a B.A. in 1930 and an M.A. in 1947. His Ph.D. in 1935 was from Harvard. He was Sterling Fellow at Yale in 1935–1936 and Benjamin Peirce Instructor at Harvard in 1936–1938. During the interval 1933–1938 he was secondarily a partner in the Dale Richardson Laboratories, where a principal product was prepared dogfish for dissection in school laboratories and which operated on the island where he had a summer home. From 1938–1947 he was associate professor and professor at Kenyon College. During several of those years he was on leave as scientific liaison officer in the Office of Scientific Research and Development in London, 1944–1946, and scientific director in 1946–1948. He spent more than a year as chief of the fundamental research branch of the Atomic Energy Commission. He then became executive director of the American Mathematical Society in November 1949, where he served until 1954. He was professor and chairman of the Department of Mathematics at Washington University in 1954–1961 and then professor at Case Institute of Technology (later part of Case Western Reserve University) from 1961. He died on 30 September 1973.

John Hamilton Curtiss

John Hamilton Curtiss was born in Evanston, IL on 23 December 1909. His father was the mathematician D. R. Curtiss, then on the faculty of Northwestern University. He took his A.B. at Northwestern in 1930, his S.M. from the University of Iowa in 1931, and his Ph.D. from Harvard in 1935. He taught at Cornell until 1943, when he went to the National Bureau of Standards. In 1944 he became chief of the Applied Mathematics Division. He was awarded the Meritorious Service Medal of the U.S. Department of Commerce in 1949. He was at the Courant Institute in 1953–1954. He became executive director of the American Mathematical Society in 1954 and remained there until 1959. He then became professor of Mathematics at the University of Miami, and served as chairman for two years. He retired in 1975 and died on 13 August 1977.

Gordon Loftis Walker

Gordon Loftis Walker was born in Salt Lake City, UT on 29 October 1912. He received a B.S. from Louisiana State University in 1937 and an M.A. in 1938. His Ph.D. was from Cornell in 1942. He taught at the University of Delaware, Temple University, and Purdue University before becoming head of the mathematics section of the Research Center of the American Optical Company in 1954. He moved to the American Mathematical Society as executive director in 1959 and remained in that position until his retirement in 1977. He continued thereafter as a consultant to the Society working in the programming of substantial tasks in the computerization of Society procedures.

William Judson LeVeque

William Judson LeVeque was born in Boulder, CO on 9 August 1923. He received his B.A. degree from the University of Colorado in 1944. His M.A. was from Cornell University in 1945 and his Ph.D. from the same institution in 1947. He was Benjamin Peirce Instructor at Harvard in 1947–1949 and then went to the University of Michigan where he rose to the rank of professor and was chairman of the department in 1967–1970. From 1970 to 1977 he was professor of Mathematics at the Claremont Graduate School. In 1977 he became executive director of the American Mathematical Society, a post he held until his retirement in 1988. This was not his first post with the Society. In particular, he was executive editor of *Mathematical Reviews* during 1965–1966 while on leave from the University of Michigan.

With the anticipated retirement of LeVeque about August 1988, a search committee was established in 1986, consisting of F. W. Gehring, chairman, G. D. Mostow, F. P. Peterson, E. Pitcher, and P. A. Sally. The committee gave various levels of consideration to forty candidates or potential candidates for the position of executive director. The committee recommendation, selected by the Council and approved by the Trustees, was William H. Jaco. His Ph.D. is from the University of Wisconsin under D. R. McMillan, Jr. His principal academic appointments have been at Michigan, Rice, and Oklahoma State. His most recent appointment was as Professorial Research Fellow at the University of Melbourne in 1987-1988.

The Society had some difficulty accommodating to the position of executive director. The creation of the position made it possible for the Council and the Trustees to relinquish detailed work and administrative functions that they had been performing but they did not always do so. The Trustees never could bring themselves to the position of some corporations that the Trustees function entirely by employing a chief executive officer, appraising the work, and firing if it is not satisfactory. There are at least three reasons. One was the inertia of change. A second was a wish to keep control in the hands of the amateur volunteers who are active research mathematicians. The effort to control shifts from creating policy to performing operations. Finally, the presence of *Mathematical Reviews* with a strong executive editor and a second geographic location sometimes put the Trustees in a mediating position. The fact that the operation of the office has become increasingly more technical with publication in-house and computerization of records and communication has greatly increased dependence on the executive director and the staff. They are needed for their expertise as well as to accomplish the work.

Trustees

In 1938 the Board of Trustees consisted of five persons elected for a term of two years. Neither the president nor the treasurer was a trustee.

The Board for 1937–1938 was W. B. Fite, Robert Henderson, W. R. Longley, G. W. Mullins, and R. G. D. Richardson. Although Richardson was secretary, he served on the Board as elected trustee and not ex officio. The same five persons were Trustees in 1939–1940.

In 1941–1942 the Trustees were Longley, Mullins, Marston Morse, Richardson, and Warren Weaver. Morse was president coincidentally. They continued in 1943–1944, 1945–1946.

Beginning in 1947 the Trustees were T. H. Hildebrandt, Longley, Mullins, Richardson, and Weaver. Hildebrandt was immediate past president. Weaver resigned in November 1947 and was succeeded by G. T. Whyburn. Mullins resigned in January 1948 and was succeeded by P. A. Smith.

Continuing in 1949, the Trustees were Hildebrandt, Longley, Richardson, Smith, and Whyburn. However, Richardson died in July 1949 and was replaced by B. P. Gill, the immediate past treasurer.

Effective in 1951 the bylaws were amended so that the Board consisted of five Trustees elected for a term of two years and the president and treasurer ex officio. Effective in 1956 the term of the elected Trustees was changed to five years. The associate treasurer became a member ex officio in 1972. Here is the subsequent succession of elected Trustees:

T. H. Hildebrandt	1947–1951
G. T. Whyburn	1947–1951
P. A. Smith	1948–1954
B. P. Gill	1949–1954
J. R. Kline	1951–1954
Einar Hille	1952–1955
W. T. Martin	1953–1954
G. A. Hedlund	1955–1958
Mina Rees	1955–1959
H. F. Bohnenblust	1954–1962

For many years the secretary, although not a trustee, served as secretary of the board. Richardson, who was secretary from 1921 through 1940, was also elected trustee beginning in 1937 and served into 1949. Every other secretary to date, namely Kline, Begle, and Green, has subsequently been elected trustee.

For a number of years beginning in 1951, the Board elected its chairman and secretary year by year, usually the president and the treasurer. By 1979 the convention had been established that the chairman was ordinarily the elected trustee in the fourth year of a term and the secretary the one in the second year. However, minutes of executive sessions of the Trustees in fact appear in segments over several names.

After the secretary had ceased to serve as secretary of the board, the Trustees on November 1982 approved a resolution "that the Secretary be authorized and invited to attend all meetings of the Board, including executive sessions." They noted that this conformed to practice ever since the incorporation but seemed nowhere to be stated.

The duties of the Trustees are specified in the bylaws. Section 2 of Article II reads as follows:

> Section 2. The function of the Board of Trustees shall be to receive and administer the funds of the Society, to have full legal control of its investments and properties, to make contracts, and, in general, to conduct all business affairs of the Society.

Section 3 in almost its present form was approved by the Council of 31 December 1940. The first sentence is the following:

> The Board of Trustees shall have the power to appoint a manager and such assistants and agents as may be necessary or convenient to facilitate the conduct of the affairs of the Society, and to fix the terms and conditions of their employment.

The section continues about delegation of powers without divestment of responsibility and served to formalize what was already the existing state of affairs. The words "a manager and" were deleted in 1960 in recognition of the fact that since 1948 there was a separate article authorizing the position of executive director but specifying since 1960 that the appointment is made by the Trustees with the consent of the Council. See the chapter on the executive director.

The Trustees have certain specified duties, among them the approval of Council actions in filling vacancies in Council membership or editorial committees, the filling of its own vacancies, and the approval of the dues and privileges of members set by the Council.

There is risk of liability for one's well intended actions as an officer and particularly as a trustee. This has become more apparent in current society with an increase in suits against directors and officers. It was in 1983 that the Society first took out directors' and officers' liability insurance. The price of this became high and there was some risk that it would become unavailable, so the bylaws were amended in 1987 to provide indemnification for expenses of defense or settlement of actions except in case of negligence or misconduct.

The work of the Board of Trustees has changed with time. In the late 1930s the Board gave direct consideration not only to such questions as the salaries of individual employees but also to publishing contracts, reprinting decisions, and to minute investment decisions such as whether to exercise or sell the right to purchase a share or two of stock or to establish a checking account. The advent of an executive director in 1949 relieved not only the secretary and the associate secretary for Financial Affairs (the office was then eliminated) of a burden of administration but also the Trustees. An investment committee to expedite the handling of investments was established in 1955. A fiscal manager appeared in 1960.

Despite the fact that the staff became larger and more professional and that the work shifted from the Trustees to the staff, there has been a tendency, recognized and resisted by the Trustees but nonetheless persistent, to "micromanage" the affairs of the Society.

Planning and Organization

The business of the Society is conducted in layers that have grown on both sides of a central structure. Originally business was done in the Council and carried out by officers. The treasurer was a business agent of the Society and the secretary conducted business activities as well. Specific recurring pieces of business, such as printing contracts, were negotiated by editors. With the incorporation in 1923 some business questions were separated from the Council. However, the Council has always impinged on financial questions in some of its policy recommendations, which may be formulated as decisions but become effective when honored by the Trustees.

The advent of the executive director in 1948 relieved the secretary and the treasurer of a great deal of the business of operating the Society. The growth of the headquarters office constitutes the layer, on one side, of the Council and Trustees, taking care of administrative functions once directly performed by elected personnel.

On the other side there has grown a committee structure. There were ad hoc and standing committees from the beginning to deal with specific policy matters. It is the standing committee structure to predigest business that goes to the Council and Trustees or to act instead of the Council or Trustees that is one of the subjects of this chapter.

THE EXECUTIVE COMMITTEE

An ad hoc Committee on Reorganization made a variety of recommendations in April 1948. One of these was for an Executive Committee along the lines of the one that now exists. The ad hoc committee was discharged but a Committee on the Bylaws was authorized. In September 1948 the proposed revision, including a new Article V on the Executive Committee as now constituted, was presented. With minor modification the amended bylaws became effective on 1 January 1950.

The Executive Committee consists of the president, the president-elect or ex-president, the secretary, and four elected members. Through 1983, the term of elected members was two years with two elected each year. Beginning in 1984 it was four years, with one elected each year. The term begins nominally on 1 January. In fact the election process cannot begin until the results of election to the Council are known (about 20 November). It has been a two-step process for many years, with a straw ballot to inform a nominating committee, followed by the election. Thus the new member may not be known until the middle of February and the departing member serves until replaced.

Principal duties of the Executive Committee are defined in Article V, Section 2 of the bylaws. It is empowered to act on matters delegated to it by the Council. It may refer matters to the Council if so requested by three of its members. It is responsible to the Council and reports to the Council. It may consider the agenda for Council meetings and make recommendations to the Council.

Until 1962, the Executive Committee appears not to have been systematically active. On 23 January 1962, in response to recommendations from a Planning Committee, and again on 19 August 1975 the Council delegated duties to the Executive Committee as follows:

1. Elect members of various classes.

2. Set dates and places of meetings.

3. Approve subjects and organizing committees of symposia, institutes, etc., in established series or upon recommendation of appropriate committees.

4. Approve publication of volumes in established series, such as Colloquia, Surveys, Proceedings of Symposia, upon recommendation of the appropriate committee.

5. Approve recommendations of Editorial Committees concerning numbers of pages to be published, interim editorial appointments, etc.

6. Approve invitations to give special lectures, such as the Gibbs Lecture and Colloquium Lectures, on recommendation by the appropriate selection committees.

7. Approve winners of prizes, such as Bôcher, Cole, Veblen, Birkhoff, Wiener, and Steele prizes on recommendation by the appropriate selection committees.

8. Approve panel discussions, short courses, and other events, either peripheral to a meeting or separate from meetings, including joint sponsorships of events in which the participation is real.

9. Receive reports of committees of the Council that recommend no action.

10. Set dates and places of Council meetings.

The procedure on election of members and setting dates and places of meetings and Council meetings (items 1, 2, 10) was changed in 1983 when these duties were further delegated to a committee consisting of the Secretary and the Associate Secretaries.

The procedure on approvals (items 6 and 7) was changed in November 1984. Selection committees were asked to submit semifinal lists of candidates for invitations or prizes to the Executive Committee for comment. Following the comments, the selection committee makes its choice.

The Council of 20 January 1971 received a long report of an ad hoc Committee to Review Society Activities, on which it acted. The fourteenth and last recommendation, approved by the Council "charged [the Executive Committee] to carry out a continuing review of Society activities." The Executive Committee agreed to carry out this charge in a three-year cycle, looking at meetings of all kinds, publications, and all other activities in turn. For several cycles, the process consisted of a draft message on the state of affairs prepared by the Secretary, revised by the Committee, and presented to the Council. This was ultimately regarded as a not sufficiently penetrating review. In 1986, which was a year to examine the residual set, each member of the Executive Committee prepared a critique of a phase of activity. These were discussed and summarized with the assistance of Allyn Jackson, the newly appointed staff writer. With further review this became the report to the Council.

Other duties have been assigned to the Executive Committee. In approving the subject matter for grant proposals for conferences, it is specifically charged with preventing duplication of covering of fields in instances where that might be undesirable.

The Executive Committee has noted that the bylaws contain two provisions that might come into conflict.

Article V, Section 2, says in part:

> If three members of the Executive Committee request that any matter be referred to the Council, the matter shall be so referred.

Article V, Section 3, says in part:

> An affirmative vote on any proposal before the Executive Committee shall be declared if, and only if, at least four affirmative votes are cast for the approval.

Inasmuch as there are seven members on the EC, it is clearly possible that both positions might be cited over a single issue and might raise the question of which should prevail. To avoid this dilemma, the Executive Committee adopted the following rule:

When the Executive Committee is considering whether to act on a proposal under Article V, Section 3, of the bylaws or to refer it to the Council under Article V, Section 2, the latter issue shall be resolved first.

The first joint meeting of the Executive Committee with the Board of Trustees appears to have been in May 1963 at the invitation of the latter. From that time on, joint meetings have been the regular arrangement. Over a long period of time, it is observed that several kinds of business have moved from the Council to the Executive Committee. This follows not only from the delegation of powers but also from increasing use of the power to report actions to the Council and to make recommendations to the Council. It is convenient to consider the scientific content of an issue and the financial implications at a joint meeting of the Executive Committee and the Board of Trustees. If this shift in the locus of some categories of business were not to take place, the meetings of the Council would be intolerably protracted and the Council would have no time for the things that it still does.

EXECUTIVE COMMITTEE

Presidents (ex officio)

E. Hille	1949	O. Zariski	1968–1971
J. L. Walsh	1949–1951	N. Jacobson	1970–1973
J. von Neumann	1950–1953	S. Mac Lane	1972–1975
G. T. Whyburn	1952–1955	L. Bers	1974–1977
R. L. Wilder	1954–1957	R H Bing	1976–1979
R. Brauer	1956–1959	P. D. Lax	1978–1981
E. J. McShane	1958–1961	A. M. Gleason	1980–1983
D. Montgomery	1960–1963	J. B. Robinson	1982–1985
J. L. Doob	1962–1965	I. Kaplansky	1984–1987
A. A. Albert	1964–1967	G. D. Mostow	1986–
C. B. Morrey, Jr.	1966–1969	W. Browder	1988–

Secretaries (ex officio)

J. R. Kline	1949–1950	J. W. Green	1957–1966
E. G. Begle	1951–1956	E. Pitcher	1967–1988

Elected members

S. Eilenberg	1949	M. H. Protter	1967–1968
T. H. Hildebrandt	1949	R. C. Buck	1968–1969
S. Mac Lane	1949–1950	I. Herstein	1968–1969
W. T. Martin	1949–1950	V. L. Klee, Jr.	1969–1970
E. Hille	1950–1951	J. T. Schwartz	1969–1970
A. W. Tucker	1950–1951	R. H. Bott	1970–1971
W. Feller	1951–1952	C. C. Moore	1970–1971
M. H. Heins	1951–1952	M. E. Rudin	1971–1972
G. A. Hedlund	1952–1953	E. M. Stein	1971–1972
R. L. Wilder	1952–1953	W. Browder	1972–1973
E. E. Moise	1953–1954	F. W. Gehring	1962–1973
J. C. Oxtoby	1953–1954	P. R. Halmos	1973–1974
J. L. Doob	1954–1955	D. M. Stone	1973–1974
G. B. Price	1954–1955	C. S. Morawetz	1974–1975
D. Montgomery	1955–1956	I. Singer	1974–1975
C. V. Widder	1955–1956	C. W. Curtis	1975–1976
K. D. Friedrichs	1956–1957	J. T. Schwartz	1975–1976
A. Zygmund	1956–1957	R. G. Bartle	1976–1977
W. S. Massey	1957–1958	R. C. Kirby	1976–1977
A. H. Taub	1957–1958	W. Browder	1977–1978
R H Bing	1958–1959	I. Singer	1977–1978
E. Hille	1958–1959	L. P. Rothschild	1978–1979
A. M. Gleason	1959–1960	S. S. Shatz	1978–1979
R. P. Halmos	1959–1960	F. E. Browder	1979–1980
L. Bers	1960–1961	R. G. Douglas	1979–1980
G. P. Hochschild	1960–1961	R. L. Graham	1980–1981
R. H. Bott	1961–1962	F. W. Gehring	1980–1981
R. C. Buck	1961–1962	M. E. Rudin	1981–1982
M. M. Day	1962–1963	I. Singer	1981–1982
G. A. Hedlund	1962–1963	P. R. Halmos	1982–1985
P. D. Lax	1963–1964	P. J. Sally, Jr.	1982–1983
L. Zippin	1963–1964	E. M. Stein	1983–1984
P. R. Halmos	1964–1965	M. Hochster	1983–1986
L. Nirenberg	1964–1965	H. Bass	1984–1987
A. Rosenberg	1965–1966	J. E. Taylor	1985–1988
E. H. Spanier	1965–1966	W. A. Veech	1986–
N. E. Steenrod	1966–1967	I. Kra	1986–
D. Zelinsky	1966–1967		
G. W. Mackey	1967–1968		

Liaison and Budget Committee
and
Agenda Committee

As noted in the section on the executive director, the Liaison Committee, consisting of the president, secretary, and treasurer, was formed at the time of the appointment of the second executive director to serve as a reference point for the continuing operations of the executive director. With time it served not only that purpose but also as a locus for discussion of proposed administrative decisions and for formulation of problems for the more formal bodies. As the latter role expanded it became an agenda committee for the Executive Committee and the Board of Trustees.

The budget of the Society from the beginning was in the hands of the treasurer, assisted by the office manager and later by the executive director. By 1968 there was a Finance Committee of the Trustees, of which the treasurer was a member. Beginning in 1974 the Liaison and Finance Committees met jointly as a Budget Committee to review the budget as prepared by the executive director and staff before the Trustees examined and modified and approved it.

The Liaison Committee maintained its role as reference point for the executive director. In its other functions it developed with the Budget Committee into the Agenda and Budget Committee (ABC), formally constituted in 1981. Beginning in 1982 the ABC met regularly a month or so before each joint meeting of the Executive Committee and Board of Trustees (ECBT) to go over a preliminary version of the agenda including the budget, deciding some minor points, placing proposed decisions on a consent agenda, modifying presentations, and making recommendations. The intent was to cut the bulk of material otherwise handled by the ECBT by doing some of the work and organizing the rest.

The increased volume of business and of staff support and the systemization of handling business in layers removed some kinds of business from the Council and the Trustees so that minutes of these bodies have changed markedly over the fifty-year interval. The Trustees are no longer concerned with salaries of individual staff members except for a small subset or with specific investments and the Council no longer considers whether to accept a specific book manuscript or how many pages each journal should publish in a year. As a consequence however some members of the Council feel separated from the process of making decisions.

Planning Committee

The Society has had many committees to review a segment of activity or to plan. Some of these will be reviewed here. A very influential one was the Planning Committee that was active in 1961–1962. It consisted of Richard Brauer, Saunders Mac Lane, chairman, J. L. Doob, Will Feller, E. E. Floyd, J. W. Green, A. E. Meder, Jr., and L. J. Paige.

In its first report to the Council of 29 August 1961, it recommended sectional meetings on special topics, restricting publicity at Summer and Annual Meetings to announcements of the meeting and of the invited lectures (attempts at publicity had been unsuccessful), and the routing of lecturers in the Visiting Lecturer Program to the smaller centers. The committee noted that it had considered strengthening the Executive Committee and having it meet regularly and the establishment of a Mathematics Service Organization for publishing and other business operations for both the AMS and other organizations. On these two points it made no recommendation. The Committee recommended that the Society take on the publication (but not the editorial responsibility) of *Mathematics of Computation* for a limited period. The Council approved the recommendations.

In a second report on 23 January 1962 the committee restated the purposes of the Society: in summary, these are scientific meetings for presentation of research, publication of research, recognition of achievement, and dissemination of results by reviewing and publication of monographs. The Society is also concerned with secondary means of support of its objectives. The recommendations covered these points. First was that the Executive Committee should meet three times a year before Council meetings with an assigned list of delegated duties and the duty of preparing and predigesting the Council agenda. This was approved, including a list of duties, as detailed elsewhere. Second was an experimental approach to Summer and Annual Meetings (including the idea of Special Sessions). This was approved on a limited basis. Third was the establishment of an advisory committee on issues related to the federal government. This was withdrawn. Fourth was the reinstatement of the column of "Letters to the Editor" in the *Notices*, which had been suspended. This was approved with the understanding that there be editorial selection and supervision. At the same time the Secretary was made an editor of the *Notices* along with the executive director. For the most part these recommendations have been discussed elsewhere in this book. Although the details of execution may have differed from the proposals, the influence of the effort was substantial.

THE SPECIAL REVIEW COMMITTEE

The Trustees become concerned from time to time about whether they are performing their duties effectively. This concern is exacerbated by the increasing span and bulk of the enterprise that they supervise. The change is epitomized in the amount of paper that they see. In 1938 for the entire year there were 10 pages of minutes of the Trustees and 36 pages of supporting accounts. By 1987 the minutes of one of two meetings of the Executive Committee and Board of Trustees came to 23 pages (plus pages of an executive session) with 219 pages of attachments and a book of summary financial reports of 122 pages. The latter were presented only in summary because by 1982 the full set of financial reports came to 1017 pages and its distribution was discontinued.

As a consequence, the Trustees from time to time consider seeking advice. One route that they contemplate is the use of consultants on how to manage their affairs. The use of consultants in particular technical problems of the Society has been effective, for instance on problems concerned with computer equipment and operation, but on the badly posed problem of how to run Society business it was less so. The Society had one contact with a distinguished firm of consultants who pay special attention to nonprofit organizations but there was little return. It is the opinion of some Trustees, though this is an egocentric view, that mathematicians are different, that it is not possible to explain effectively to an outsider such as a consultant what the Society is about, and so it is impossible to get useful advice.

The Trustees faced the general problem of the effectiveness of their operation again in 1983 with the public resignation of Alex Rosenberg toward the end of his tenth year as Trustee over his stated lack of confidence in the Providence management, partly because of deficits that had been incurred. Rather than turn to professional consultants, the Society established a Special Review Committee consisting of Richard A. Leibler, formerly the director of the Institute for Defense Analysis in Princeton and then a consultant, and W. Ted Martin, treasurer in the interval 1965–1973 and for many years head of the mathematics department at M.I.T.

The Special Review Committee reported "that the Providence office is well managed and that it operates smoothly and efficiently." The deficits were laid to the development costs of a *Mathematical Reviews* data base, the fact that journal prices had not been raised sufficiently, and the fact that evasion of registration fees and purchase of books and journals at individual rates for libraries had not been policed.

LONG-RANGE PLANNING COMMITTEE

The Society has been heavily dependent on committees to do its work. Planning activities are a natural committee activity. In May 1984 a Long-Range Planning Committee (LRPC) was established. It was an outgrowth of the Special Review Committee. It was appointed by the Chairman of the Board, Cathleen S. Morawetz, as an advisory committee to the Executive Committee and Board of Trustees (ECBT). Its initial membership was M. Hochster, chairman, H. Bass, R. L. Graham, F. P. Peterson, and E. Pitcher. It had a very broad charge, of which the first item was "[t]o suggest a number of different management plans to improve the operation of the Society." Another was "[t]o offer suggestions for making the budget process safer."

In fact in its initial report to the ECBT of May 1985 the LRPC found sufficient consensus that only a limited number of alternative suggestions appeared. Many recommendations were approved immediately by the ECBT including a variety of procedural matters. One was to make the Agenda and Budget Committee (ABC) more of a budget committee with less of an agenda function (countervailing some of the initiative that established the ABC). A second was to make the LRPC into a standing committee. Another was to change the procedure for amendment of the bylaws so that approval by the membership could be done by mail ballot, thereby permitting broader and less selected participation by the membership. This was soon accomplished by amendment of the bylaws. There were proposals for fund raising and for improvement of public relations, of which the most forceful was to support the recruitment of young mathematicians into the profession. As to budget questions, it was agreed that a reserve fund equal to one year's budget should be created over a period of years.

The LRPC reported again to the ECBT of November 1985. The permanent structure that was proposed and adopted called for a standing but relatively inactive committee that becomes intensely active one year out of five or at the request of the president or the chairman of the Board. Its membership should consist of the secretary, the treasurer or associate treasurer, the trustee in the third year of a term, and the members of the Executive Committee in the second and third years of their terms. Recommendations of the LRPC included moderate expansion of the publication program and ways and means to support the role of applied mathematics in the programs of the Society.

One delicate issue that was discussed at this time and later was the place of *Mathematical Reviews* in the hierarchy of the Society. These discussions resulted in the affirmation noted in the chapter on *Mathematical Reviews* that the executive director is the chief executive officer of the Society, including *Mathematical Reviews*.

The Council of 23 April 1988 at the urging of W. P. Thurston established a committee consisting of R. M. Fossum, A. Marden, M. Pour-El, chairman, and H. Stark to reconsider the procedures of the Council for considering issues, and to review the set of powers delegated to the Executive Committee.

Policy Committees

There has been a succession of joint policy committees in the profession. The first was the War Policy Committee of WWII, which evolved into the Policy Committee. The subsequent developments were in two directions. One was to a more formal successor, the Conference Board. As it developed, it took care of only a fraction of the joint needs of the research community and a Joint Policy Committee was formed and evolved into the Joint Policy Board for Mathematics.

THE WAR POLICY COMMITTEE

World War II had a profound effect on American mathematics. One aspect dealt with in another chapter, was the increased prominence of various aspects of applied mathematics. Moreover, there were effects on organization. The immediate effect was the War Policy Committee but successor organizations lived on after the war was over.

The Council of 7 September 1939 empowered the president to appoint a committee, jointly with the Mathematical Association of America, "to advise regarding mathematics in preparedness measures, teaching, research, etc." The president was requested "to communicate with the Division of Physical Sciences of the National Research Council to express the willingness of the Society to cooperate and to request advice as to what further steps should be taken at this time." President G. C. Evans and President W. B. Carver of the MAA did appoint a Preparedness Committee. The general chairman was Marston Morse. There were two subcommittees:

Research Subcommittee: Dunham Jackson, chairman, Harry Bateman, E. J. McShane, M. H. Stone, E. B. VanVleck, Norbert Wiener, S. S. Wilks

Instruction Subcommittee: J. L. Coolidge, chairman, L. R. Ford, E. C. Goldsworthy, W. L. Hart, M. H. Ingraham, E. J. Moulton, D. V. Widder.

A study of the special function of mathematicians in World War I was undertaken and a census of available mathematicians was considered.

Professor Morse reported for the War Preparedness Committee at a joint meeting of the Council and the Board of Governors of the MAA on 9 September 1940. The text of the report is in the *Bulletin*, v. 46 (1940), pp. 711–714. The committee stated three objectives, quoted here:

1. The solution of mathematical problems essential for military or naval science or rearmament.

2. The preparation of mathematicians for research essential for objective (1).

3. The strengthening of undergraduate mathematical education in our colleges to the point where it affords adequate preparation in mathematics for military and naval services of any nature. A study by a large group of mathematicians of the current routine military texts and sources where mathematics is involved in order that mathematicians may exert their proper influence on the teaching of military and naval science in time of war.

The Research Committee already appointed was continued as the Subcommittee on Research. The Instruction Subcommittee was replaced by two committees as follows:

Subcommittee on Preparation for Research: M. H. Stone, chairman, B. O. Koopman, R. E. Langer, Hans Lewy, F. D. Murnaghan, H. P. Robertson

Subcommittee on Education for Service: W. L. Hart, chairman, R. S. Burington, J. L. Coolidge, H. B. Curry. E. C. Goldsworthy, F. L. Griffin, M. H. Ingraham, E. J. Moulton.

The consultants were:

Aeronautics: Harry Bateman, chief; Ballistics: John von Neumann, chief, W. T. Reid; Computation: Norbert Wiener, chief; Cryptanalysis: H. T. Engstrom, chief, A. A. Albert, W. A. Hurwitz, Solomon Kullback, Øystein Ore; Industry: T. C. Fry, chief; Probability and Statistics: S. S. Wilks, chief.

Other subcommittees were appointed.

At a joint meeting of the Council of AMS and the Board of Governors of MAA on 27 December 1942, the War Preparedness Committee was discharged at the suggestion of its chairman. The reason is not immediately apparent in either the minutes of the Council or the files of Secretary J. R. Kline, though of course the name War Preparedness Committee was no longer appropriate. In its place, a joint War Policy Committee was appointed. The committee consisted of the following persons:

M. H. Stone, chairman, W. D. Cairns, G. C. Evans, L. H. Graves, Marston Morse, Warren Weaver, G. T. Whyburn.

Moreover, Stone was requested to represent mathematicians in Washington until the committee was functioning. He actually continued in that role for an extended period.

There were changes in personnel of the committee from time to time, some caused by wartime duties of members that conflicted with the committee service.

The report of C. C. MacDuffee, acting chairman of the War Policy Committee, dated 1 September 1943, shows that the Rockefeller Foundation supported the committee with a grant of $2500. There were subsequent grants of $2500 and $1000.

The War Policy Committee was organized with three subcommittees.

> Subcommittee on Available Teachers of Collegiate Mathematics: J. R. Kline, W. D. Cairns, Arnold Dresden
>
> Subcommittee on War Training Programs: W. L. Hart, chairman, C. R. Adams, H. M. Bacon, Ralph Beatley, B. H. Brown, H. J. Ettlinger, C. V. Newsom, W. M. Whyburn
>
> Subcommittee to Advise Examinations Staff of the United States Armed Forces Institute: W. T. Reid, chairman, Ralph Beatley, L. L. Dines, W. L. Hart, C. C. MacDuffee.

Some of the activities of the War Policy Committee can be found in its reports to the Society and the Association. One of the first actions, approved by the Council of 27 February 1943, was a recommendation that the Society and the Association "continue to hold mathematical meetings, emphasizing the importance of sectional meetings, and arranging national and sectional meetings at such times and places as will put the least burden on transportation facilities."

The sentence in Article VII, Section 1, of the bylaws requiring that the Annual Meeting be held between 15 December and 15 January was suspended for the duration. Thus the Annual Meeting of 1943 was held on 26–27 November in Chicago. Later in the war there were enforced cancellations of meetings because of restrictions on transportation. Papers contributed in person were handled by title.

A recurring issue, even from the days of the War Preparedness Committee, was draft deferment. The position adopted was to request occupational deferment for all teachers of mathematics in accredited colleges and universities and of graduate students of mathematics preparing for research or college teaching. Changing regulations required continuing examination of the situation. Dissemination of regulations was undertaken.

Dues for persons in the armed forces were set at $1.00.

There was great demand for teachers for programs of the armed forces in colleges and universities. As the draft brought personnel into the ranks faster than they could be deployed effectively, the training programs were used for the double purpose of improving the effectiveness of the draftees and as a "parking" place for them.

The advice to the Armed Forces Institute consisted in part of supplying examinations for accreditation purposes and transfer credit.

The committee was very much concerned with arranging that mathematicians in the armed forces were used in places where their expertise was of value. Thus they sought areas in the armed forces where mathematics was needed and tried to keep track of mathematicians in uniform to arrange to match persons with assignments.

The effective use of mathematics was also approached through the consultants and through the matching of senior mathematicians with tasks in their fields of expertise.

The report of MacDuffee already mentioned emphasized the desirability of preserving a record about the war activities of mathematicians. It notes the difficulty of preparing a current record both because some of the work was confidential and because no one was available to do it. So far as the writer is aware, the appeal that this record be assembled later has not been properly answered although some segments have been covered. The Research and Development Division of the War Department conducted a study. A questionnaire was distributed through the Institute for Mathematical Statistics and the Society asked individuals for a resumé of activities in the interval 1942–1946, particularly the degree to which their professional training was used by the armed services. Physicists and chemists were questioned similarly.

As soon as the war was over, the War Policy Committee was discharged. The formal action took place at the Council of 16 September 1945. The letter from M. H. Stone, the chairman, presented as a final report, recommending the discharge also recommended that a joint policy committee among mathematical organizations be established. The report was approved.

The Policy Committee

Following the recommendation of the War Policy Committee as it was discharged, the Council of 23 November 1945 considered a plan for a Policy Committee for Mathematics as prepared by President T. H. Hildebrandt and Secretary J. R. Kline. With slight modification by the Council it was approved in the following form:

1. There shall be established a Policy Committee for Mathematics, to be composed of four representatives from the American Mathematical Society, two from the Mathematical Association of America, one from the Institute of Mathematical Statistics, and one from the Association for Symbolic Logic. The committee shall elect its own chairman.

2. Representatives of each organization shall be selected in accordance with a plan approved by the governing body of that organization.

3. The Secretary of the American Mathematical Society shall be a non-voting, ex officio member of the committee and shall act as secretary for the committee.

4. The Policy Committee shall study those problems affecting the mathematical profession which are the common concern of the constituent organizations. It shall be empowered to speak for the constituent organizations on matters which concern the position of mathematics in such matters as proposed or enacted legislation concerned with science, problems concerning the effective use of mathematicians or potential members of our profession, and other questions which tend to affect the dignity and the effective position of mathematics among related sciences, both nationally and internationally.

4. Nothing in the powers of this committee shall be construed to affect any commitments already made on a national or international basis by any of the constituent organizations (i.e., among these is the International Congress of Mathematicians for which an invitation was issued by the American Mathematical Society in 1936).

5. The work of this committee shall be financed by one of the following methods:

   (1) By means of funds requested from national foundations for this specific purpose;

   (2) By means of funds contributed by the constituent organizations, in proportion to their voting representation on the committee.

6. This Policy Committee shall be approved for a period of five years. At the end of that time the work of the committee shall be reviewed and a decision made concerning the continuation of the committee.

The Council agreed that a nominating committee selected by the president should name eight candidates for the four Society places on the Policy Committee. The nominating committee, consisting of M. H. Ingraham, chairman, J. R. Kline, and R. L. Wilder named A. A. Albert, G. C. Evans, T. H. Hildebrandt, R. E. Langer, E. J. McShane, Marston Morse, G. B. Price, and M. H. Stone. Those elected were Stone (term four years), Morse (three years), Hildebrandt (two years), Evans (one year) in order of number of votes received.

The Institute for Mathematical Statistics and the Association for Symbolic Logic were represented immediately by Will Feller and Alonzo Church respectively. However, the Mathematical Association tabled the matter of participation on initial consideration. In 1947 the number of MAA representatives was increased to three and the MAA did join. The National Council of Teachers of Mathematics became a member with one representative effective in 1951.

The Rockefeller Foundation supported the Policy committee by extending the period of the final grant for the War Policy Committee (more than $300 remained) and making an additional grant of $1500.

The Policy Committee initially concerned itself with various bills before Congress, atomic research as it affects freedom of investigation, selective service regulations, and problems of international cooperation as raised by the United Nations Educational, Scientific, and Cultural Organization, then in the process of formation. Chief among the legislative matters was the proposal for a National Science Foundation.

The Policy Committee supported the Office of Scientific Personnel, passing through grants from the Rockefeller Foundation of $1500 in 1945–1946 and $2000 in 1946–1947.

In the election of 1946, Langer was elected to the Policy Committee over McShane as the replacement for Evans. In 1947 Einar Hille was elected to replace Hildebrandt.

The Policy Committee concerned itself with the reestablishment of the International Mathematical Union, which had expired prior to the Congress of 1936. An initial attempt to get the International Council of Scientific Unions to sponsor a discussion meeting in 1947 failed. The Policy Committee accordingly set a time just prior to the Congress of 1950 for an organization meeting. Worldwide representation on the Union, which had been absent in the defunct predecessor, was thought desirable. The Policy Committee wished to establish the principle that Congresses are independent of a mathematical union and was prepared to insist on this point if necessary with respect to the Congress of 1950, which the Society had well in hand and which was based on the previously accepted invitation to conduct the deferred Congress of 1940. It developed that both the Policy Committee and the Organizing Committee for the Congress had made overtures toward reestablishing the Union. The Council clarified that the Policy Committee was the sole agent.

The stance of the Union with respect to International Congresses was subsequently formulated by the Policy Committee on 27 December 1948 in the following principle as a basis for negotiation by Stone:

> To cooperate with the four-year Congresses when requested and
> to assist in organizing other international meetings of limited size
> and scope.

The Kilgore-Magnuson bill, proposing a version of the National Science Foundation, was supported by the Policy Committee in 1946.

Concerns of the committee in 1949 included the following topics: the possibility of a foundation or a "super-structure" to secure support from industry; selective service regulations; liaison with the Department of the Army and the Department of the Air Force at their request; the structure of the International Mathematical Union; and again the proposal for the National Science Foundation. In particular, the Policy Committee formally supported the establishment of the Foundation. A subcommittee consisting of W. T. Martin, chairman, G. C. Evans, Saunders Mac Lane, and Mina Rees was studying the way in which the National Science Foundation, if established, could serve science and mathematics. The subcommittee posed eight questions about grants, conferences, publication, and how the Foundation should inform itself about performance of grant recipients. The subcommittee suggested that governing bodies, such as the Council, might discuss the nature and propriety of action about an amendment requiring an affidavit from scholarship recipients about "membership in organizations that believe or teach the overthrow of the United States Government by force or violence, etc."

At the same time that the Council supported the establishment of the National Science Foundation, it "deplores the amendment inserting a new section ... which introduces non-scientific objectives."

The Policy Committee encouraged the formation of the Division of Mathematics in the National Research Council. This was done in 1951 with M. Morse as chairman and M. H. Stone as vice chairman. At the same time the Policy Committee was able to announce that the United States would be represented in the newly reorganized International Mathematical Union by a United States National Committee for Mathematics with five members named by the Policy Committee and four by the new Division of the NRC.

In 1951 the Council and the Board of Governors of the MAA jointly urged that the Congress appropriate funds to the National Science Foundation for the support of both basic scientific research and students taking scientific training.

In 1953 it was noted that a committee had been appointed to evaluate the functions and operations of the Bureau of Standards and that the Policy Committee would be asked to name a member.

The manner of selecting Society representatives to the Policy Committee was considered in April 1955. It was observed that although the Policy

Committee met rarely it considered items of extreme importance. Thus the Society representatives should be well-informed. It was agreed that the representatives should be the president, the president elect or ex-president, the secretary, and one of the vice-presidents named by the president.

The Conference Board

Whereas the War Policy Committee had been a loose ad hoc confederation, the Policy Committee had taken on a structured existence. In the next step, a formal framework with a constitution, bylaws, and a staff came into being. At its meeting of 27 August 1957, the Council received the proposed plan for the successor organization called the Mathematics Conference Organization. Initially its membership was the same as that of the Policy Committee. Its objectives were as follows:

> The object of the Mathematics Conference Organization shall be the coordination of the activities of its member organizations in the advancement of mathematics and its applications; in raising funds for the support of their activities; in giving advice and counsel on mathematical matters; in the promotion of research, the betterment of instruction, the training of mathematicians, and the designing of programs to support their activities; and in acting in any capacity in which the member organizations can function more efficiently as a group than separately.

The new organization was to be managed by a Mathematics Conference Board of representatives and members-at-large. It was to be financed by dues levied on the member organizations in proportion to their own total dues of individual members.

It was just as this change was under consideration that in 1957 the Society for Industrial and Applied Mathematics became a member of the Policy Committee and, of course, of the successor.

The Society ratified the constitution of the Mathematics Conference Organization at its meeting of 28 January 1958. That constitution differed from the preliminary version that the Council had seen earlier principally in that the dues were set on a less ambitious scale. The plan was to collect in stated proportions the amount necessary each year to bring a standing fund up to a total of $1250. The total and the proportions were subject to periodic adjustment.

It was not intended to trace the history of the Conference Board in detail here. Its name changed to the Conference Board of Mathematics and then to the Conference Board of the Mathematical Sciences as the number and diversity of its membership increased. The number of organizations that adhered increased to a dozen. Initially G. Baley Price was the executive secretary. He was followed in that capacity by Leon W. Cohen. In 1965, Thomas L. Saaty was employed almost full time as executive secretary. By 1968 he had been replaced by Truman A. Botts as full-time executive director. He held the position until his retirement in 1982. At this point the organization was finding it difficult to finance its operations at the scale that grants had once made possible. The level of activity and the staff were cut back. Marcia P. Sward, associate director of the MAA, held the post of administrative officer as a small fraction of her work load until she took the position of executive director of the Mathematical Sciences Education Board in 1985. Then the position was held by Louise Raphael of Howard University, followed by Peter L. Renz, associate director of MAA.

The merits of belonging to the Conference Board were examined in 1979. They included the sponsorship of the Joint Projects Committee, the Congressional Fellow program, the Washington contact through it. executive director, the naming through CBMS of members of the U.S. National Committee for Mathematics, and the CBMS Newsletter.

One influential activity of the Conference Board deserves special mention. This is the CBMS Regional Conferences. These were one-week conferences, several each year, with a principal lecturer who gave usually ten expository lectures on a field of research. The lecturer was encouraged (in particular by a manuscript fee) to prepare the lectures for publication. By 1987 the Society had published 68 of these sets of lectures.

THE COUNCIL OF SCIENTIFIC SOCIETY PRESIDENTS

The Council of Scientific Society Presidents (CSSP) was formed in 1973. The driving personality was Alan C. Nixon, who had been elected president of the American Chemical Society as an independent candidate. The organization is a "club" of presidents, not an umbrella organization of societies. The AMS has participated since the organization meeting in June 1973, when W. J. LeVeque attended in place of President S. Mac Lane.

The CSSP has been influential in Washington in making contacts with members of Congress and highly placed government administrators and in stating positions on pending legislation and appearing to defend them before Congress.

Mathematicians have been prominent in the organization. R. D. Anderson, A. M. Gleason, and Lynn Steen have served as chairman of the organization.

THE JOINT POLICY BOARD

In 1973, the Executive Committee of the Council and the Executive and Finance Committees of MAA authorized a joint committee consisting of three members of AMS, three of MAA, and two observers from SIAM to study the possibilities of establishing one overall organization in American mathematics. The members were Saunders Mac Lane (AMS), chairman, Dorothy L. Bernstein (MAA), Ralph P. Boas (MAA), Richard C. DiPrima (SIAM), C.-C. Lin (SIAM), Alex Rosenberg (MAA), P. Emery Thomas (AMS), and Gail S. Young (AMS). After considerable deliberation, the committee decided not to recommend an overall organization but to propose an explicit way of working on problems of common concern.

The report, presented to the Executive Committee of 1–2 December 1973, provided a structure and charge for the Joint Projects Committee in Mathematics. It went with ECBT approval to the Council of 14 January 1974, where it was approved with slight modification. Some of the provisions were as follows:

> JPCM will examine the relation of the development of mathematics, pure and applied, to the welfare of society and of the general scientific community, labor, industry, government, and other interested groups.
>
> ... the JPCM shall give due attention to the employment problem and keep abreast of new developments in the relationship between mathematics and society.
>
> ... JPCM may establish committees to work on specific problems and may solicit funds to support its projects.
>
> ... JPCM may employ an executive director and appropriate staff.
>
> The JPCM shall be established for a period of five years. After that it will go out of existence unless specifically renewed by the parent organization.

SIAM soon became a full member of the consortium. Each organization supplied a small set of appointed representatives.

The JPCM needed an entity to hold funds and contracts and for those purposes turned to the Conference Board of the Mathematical Sciences.

One substantial project of JPCM was the production of a volume of essays titled *Mathematics Today*. It was produced through a steering committee with J. T. Schwartz as chairman and with Lynn Steen as editor. Somewhat to the chagrin of some of the officers of the Society, it was published in 1978 by Springer Verlag rather than by the Society.

Other projects in which JPCM participated were studies of the economic status of the profession and the role of mathematics in government activities. The latter was the beginning of the idea of the "Washington presence." The JPCM participated in the entry of the AMS into the congressional science fellowship program, of which more later. The JPCM instituted panel discussion on topics of interest to the profession at joint Annual and Summer Meetings.

The life of JPCM was extended for a second five years in 1980.

In 1982, the JPCM was reorganized as the Joint Concerns Committee for Mathematics, this time with no time limit. The change was coincident with a shrinking of the Conference Board, which no longer maintained a full-time executive director and was no longer in a good position to give service to JCCM. The AMS became the holder of funds. Each organization was now represented by its president, its executive director, and a third person.

In 1983, the JCCM established the "Washington presence." The intent was to have "a knowledgeable mathematician on a part-time salary who would gather information and espouse the cause of mathematics as liaison between the mathematical community and government. This person would represent the mathematical community by developing contacts with heads of mathematical science programs in government agencies, White House Administrators, and legislators and their aides in order to present a consistent picture of the needs, policies, priorities, and accomplishments of mathematical scientists."

The three constituent organizations approved the plan. However, there was opposition to it in the Council of the AMS on the grounds that a person remaining in the position for a long time might gather too much power or might be co-opted by the governmental establishment. This argument was denied by proponents who claimed that power would not accrue but rather that familiarity and channels of information were sought and that long service was required to develop them. Accordingly, the AMS approval carried the stipulation that no person serve longer than four years as head of the office.

The JCCM was fortunate in several respects in securing the services of Kenneth M. Hoffman as director of the operation. In addition to high quality of performance, he was already familiar with the ground in that he was the executive director of the Ad Hoc Committee of the National Research Committee that produced the David Report, formally titled *Renewing U.S. Mathematics: Critical Resource for the Future*. (See the *Notices*, v. 31 (1984), pp. 434–469 and 570–616.) Moreover, MIT allowed him full time to work on the project while continuing to pay a portion of a salary.

Several major projects occurred during his tenure in which his services were effective. These included ICM-86 and the planning for the Centennial of the AMS. He was able to make the latter part of an integrated year of celebration of 100 Years of American Mathematics.

The JPCM (and later the JPBM) participated in the support of a Congressional Science Fellow. This is a program in which scientific and professional organizations fund year-long appointments of a person to serve on the staff of a member of Congress or of a congressional committee. It has advantages for the organization both in the current contact and in the accumulation of persons in the organization with knowledge of government operations. It is advantageous for the individual in broadening of background and training. It is more useful in sciences with immediate application and in areas subject to government regulation than in mathematics. The fellow in mathematics was supported through contributions by AMS, MAA, and SIAM through CBMS to AAAS, which handled orientation, placement, and payments. The mathematical community supported one fellow per year, with gaps when there were no well qualified applicants, from 1977 through 1986.

In 1984 the name of the organization was changed to the Joint Policy Board for Mathematics. The Washington Presence became known as the Office of Governmental and Public Affairs (OGPA).

The issue of the tenure of the head was raised in 1987 as the four years drew to a close. The Society suspended the four-year limit and called for a study by the members of JPBM of the appropriate length of term. The administrative supervision of JPBM had been lodged with a Directors' Committee consisting of the three executive directors, to whom JPBM assigned the task. The Directors' Committee opted for regular reexamination and reevaluation of performance and long service for one who is performing well and recommended that the head of OGPA serve at the pleasure of JPBM under a renewable term contract that gave both sides substantial notice of nonrenewal.

The relationship of the professional mathematical community with the funding of research by various arms of the government, particularly the Department of Defense, was a substantial issue, as has been noted in the chapter on political and social questions. The JPBM was involved in that the Head of OGPA, among other activities, dealt with these agencies both to find out what were their needs and their plans for financing them and to make such information available in the mathematical community. A focus of irritation for critics was the Strategic Defense Initiative (SDI). It appeared that, having learned of the interest of the agents of SDI in some areas of mathematical research, the Head of OGPA facilitated a briefing in October 1986 through the National Academy for a group of about fifty mathematicians. This was regarded unfavorably by critics. Some opposed SDI on political grounds. Others were skeptical of the technical feasibility of SDI. Some opposed any effort that facilitated work of the Department of Defense and some of the financing of research in mathematics by any government agency except the National Science Foundation. Throughout the discussion was the underlying difference in viewpoint of the discipline of mathematics by pure mathematicians and applied mathematicians.

A committee consisting of William A. Veech, chairman, M. Susan Montgomery, Hugo Rossi, David Vogan, and Robert Williams investigated the structure and function of JPBM and considered the issue of a term for the Head of OGPA and reported to the Council of 5 January 1988. They recommended that a fixed term not be set immediately. They further recommended that a second committee report on changes in structure of such nature that the Council be better able to advise JPBM and through them the Head of OGPA on what sort of activity is appropriate.

The second committee, consisting of Jean Taylor, chairman, Robert M. Fossum, and Marc A. Rieffel, was duly appointed. It made three recommendations. First, that the Society representation should consist of the president, the executive director, and a third person elected by the Council. It was suggested that the third person be a vice-president or member-at-large, elected for a two year term not coextensive with a presidential term, with at most two terms, and serving also as an ex officio member of the Committee on Science Policy. The second recommendation was that the Committee of Executive Directors be a permanent committee concerned with details of management, particularly of OGPA, but that the executive directors be nonvoting members. JPBM should concern itself with overall direction and policy. The addition of another member from each organization concerned with financial matters is reasonable. For the Society this should be an elected Trustee. Finally, the Society through its Committee on Science Policy and otherwise should reexamine its own policy with respect to JPBM and OGPA. The reaction of the Executive Committee and Board of Trustees and of the Council to these proposals was awaited at the time of this writing.

Other Committees

The description of a committee whose function is related to an activity of the Society that is otherwise described has been presented as part of the outline of the activity. Here is an account of some committees that have wider span or more independent existence.

COMMITTEE ON HUMAN RIGHTS

The Committee on Human Rights of Mathematicians was established in April 1976. There had recently been several cases in which the Council had been interested. Among them was that of José Luis Massera, imprisoned in Uruguay for political reasons, on whose behalf President Lipman Bers had sent a protest that was endorsed by the Council. An ad hoc Committee on Principles and Procedures under the chairmanship of Edwin E. Moise proposed such a committee. The Council favored the establishment of the committee but thought that the charge and scope of the proposed committee were not clear. On the promise of President Bers to prepare a charge for consideration by the Council, the committee was authorized.

The charge appeared in August 1976. Several of its features should be noted. First, the charge as initially write concerns foreign mathematicians. President Bers later said that "Restriction to foreign mathematicians was motivated by a desire to proceed slowly and by the desire to avoid people in the States who may interpret denial of promotion, etc., as a human rights violation." It is to assist the Council through recommendations. Mathematician is broadly defined.

Violations are those enumerated in the Universal Declaration of Human Rights and the Affirmation adopted by the National Academy of Sciences, USA. Investigation accompanied by polite inquiries to foreign governments is a primary tool. The advice of other experienced organizations is to be sought. The committee is to avoid political orientation.

At the same time that the charge was approved, the Council endorsed both the Universal Declaration and the Affirmation already mentioned.

In its operation, the Committee has made polite inquiries on its own recognizance by authorization of the Council but has otherwise brought individual cases before the Council with a recommendation for a Council resolution or a presidential action with a suggested route for its dissemination.

It is not easy to measure the effect or the success of the committee. In any case in which it is involved, it is only one of many voices of concern or protest. When it becomes apparent that an instance of oppression is finally relieved, as with Massera or Anatoly Shcharansky, the Society can hope that its efforts, joined with those of many others, are partly responsible. In the instance of the annulment in 1982 of the degrees of E. P. Gil'bo, following his request to emigrate from the USSR; there was no substantive reaction to inquiries. On more general issues, such as the sporadic closing of the University of Bir Zeit in territory occupied by Israel or the policy of Berufsverbot in the Federal Republic of Germany, the Society has had no discernable effect.

When the charge was prepared it was in the minds of some that a companion charge concerning U.S. mathematicians might be prepared in a manner to exclude such problems as those of promotion and tenure but to include political persecution. In fact this did not happen. In January 1987, the charge was amended simply by removing the restriction to foreign mathematicians. In addition, a sentence is added to the clause in which violations are defined by reference to the "Universal Declaration of Human Rights" and the "Affirmation" of the National Academy of Sciences, USA. This includes the failure of government authorities to protect from attack or intimidation by private groups.

COMMITTEE ON EMPLOYMENT AND EDUCATIONAL POLICY

In 1969 a Committee on Analysis of Employment Data was appointed, consisting of William L. Duren, Jr., chairman, R. D. Anderson, and Gail Young. The impetus was a letter from Alex Rosenberg to President Oscar Zariski concerning the job market for new Ph.D.'s. The fear was that the nature of the market would force new research oriented mathematicians into institutions that were not research oriented.

Whereas the initial charge to the committee seemed to be statistical analysis of employment data, the Council appeared to envisage a larger problem of spreading information and making recommendations to improve the situation. The charge to the committee was enlarged by the Council of 20 January 1971 to include the study of the availability of jobs at all levels, alerting the community to problems relating to the employment of mathematicians, making recommendations about admission to graduate school and types of programs.

The name of the committee was changed to the Committee on Employment and Educational Policy (CEEP). John W. Jewett was added to the membership. The first report on trends and expectations in employment appeared in the *Notices* for April 1971, the Committee having been given direct access to the *Notices*.

In its report to the Council of 29 August 1971, CEEP took a more pessimistic view of the employment picture than had been exhibited in its report in the *Notices*.

The committee conducted panel discussions to gather information and to publicize the problem.

Other problems were assigned to the committee. A resolution of the Business Meeting of January 1972 had asked the Council to consider the problem of "The economic status of the profession to the colleges and universities and to society in general." Although the committee responded, the Council was not satisfied and reassigned the problem on the grounds that the committee had been constituted for a different purpose.

The committee became involved in the study of problems of teaching loads and class sizes. It urged the conversion of teaching assistantship funds into junior faculty positions.

In 1974 there was an ad hoc Committee on the Emergency Employment Situation in Mathematics (CEESIM). It was charged "to examine all aspects of this employment situation [for young mathematicians] ... and to bring to the Council proposals both for possible action by the Society and possible recommendations by the Society to other bodies." The committee soon realized that there was a substantial continuing problem and the committee was dissolved with referral of its charge to CEEP. The charge to CEEP was formally revised in January 1977 to encompass the added duty.

CEEP was charged with promoting activities that would enable mathematicians to broaden their knowledge to increase their opportunities for employment. To carry out this duty, there was established a Short Course Subcommittee, which arranged brief concentrated tuition courses at Society meetings. (See the account of Proceedings of Symposia in Applied Mathematics, the series of PSAPM.)

A formal Data Subcommittee was established for the assigned statistical problems with staff support from the Providence office. The Data Subcommittee became the source of the Salary Survey, published since 1957 in the *Notices*. There is an annual questionnaire covering not only salaries but also a variety of material about production of Ph.D.'s and employment, which is analyzed annually for publication in the *Notices*.

In January 1978 at the behest of CEEP the Council passed a resolution emphasizing that "mathematics faculty positions should be filled by professional mathematicians working under conditions conducive to their mathematical development" and deploring the systematic use of one-year appointments to fill regular positions.

In January 1981 the Council passed a resolution generated in CEEP concerning conditions and procedures related to the closing of a graduate program.

ACADEMIC FREEDOM, TENURE, AND EMPLOYMENT SECURITY

In 1971 the editors of the *Notices* were considering whether to publish a letter from Professor Dov Tamari concerning a dispute over employment with Technion (Israel Institute of Technology). The editors initially declined but later agreed to publish the letter with comments from various parties and a rejoinder by Tamari, subject to editing. While the editing was in progress, the Trustees inquired into the situation, sought legal advice, and defeated a motion to the effect that they were willing to see the letter and accompanying material published. When asked to reconsider their position by Tamari and some of his supporters, the Trustees reaffirmed their position in a positive statement. The Trustees advised the Council that publication of the letter would not be in the best interests of the Society from either a fiscal or a legal point of view.

Faced with the advice of the Trustees, the Council of 29 August 1972 authorized the president to appoint a standing Committee on Academic Freedom, Tenure, and Employment Security (CAFTES) and referred the dispute between Tamari and Technion to the committee.

The committee initially consisted of Paul S. Mostert, chairman, Paul J. Sally, Jr., and Murray Gerstenhaber. It reported in April 1975 that it had seen 19 cases, of which 4 had been settled more or less satisfactorily, 5 had been rejected as having insufficient merit, and 7 were active.

Since 1974 there had been a Committee on Legal Aid (CLA), which could certify recommended cases for financial assistance in legal problems and one case had been so recommended and certified.

It was proposed to merge CAFTES and CLA on the grounds that their problems overlapped and the former was overworked while the latter was underutilized. There was no action on this proposal but the CLA was discharged in 1982 with the understanding that the Trustees would function directly were it necessary.

There was also since 1971 a Committee on Dismissed Mathematicians "to find, whenever possible, new positions for mathematicians who have been dismissed from their positions under unusual circumstances." CAFTES was charged to cooperate with it.

The original charge to the committee had been prepared by the secretary on instruction from the Council. As to kinds of cases to consider, these might arise by referral from the Council or the Executive Committee. Cases coming from a direct appeal from individuals were to receive preliminary examination to determine whether they merited substantial consideration. The committee was not to consider cases where there is a more natural point of reference, such as a government agency or the American Association of University Professors. The services of the committee were primarily for American members of the Society who are aggrieved. The committee was to determine facts and report to the Council, possibly with a recommendation to publish some version of the report. The committee was not to be an arbitration board and was not to be a party to disputes. It could call on experts.

The charge was modified by authorization of the Council of 26 January 1977. At the Council of 21 January 1976, Karl K. Norton proposed several specific modifications of the charge. These were referred to an ad hoc Committee to Write Rules for the Operation of CAFTES, which produced the nine-page document approved by the Council of January 1977. The committee was enlarged to nine persons with staggered three-year terms.

CAFTES was involved in the case of the closing of the graduate program in mathematics at Belfer Graduate School of Yeshiva University, as a result of an appeal by faculty members. An account of the affair can be found in the *Notices*, 26, 200 (February 1979). In the course of the development, the Society agreed to serve as institutional affiliate for any member of the faculty in applying for a research grant with the National Science Foundation. This was a position that the Society had taken in other circumstances but so far as the writer knows no individual has taken advantage of this opportunity. (The NSF has maintained that no institutional affiliation is necessary, and there are instances to demonstrate the point, but individuals seem to think that it is an advantage and almost all holders of grants are affiliated.)

A summary account of the dispute between Tamari and Technion was published in the *Notices*, 30, 232–233 (February 1983).

OPPORTUNITIES IN MATHEMATICS FOR DISADVANTAGED GROUPS

In a referendum conducted in May 1969 there were six resolutions, already discussed in the chapter on political and social questions. Except for Resolution B, all were defeated. It was recognized that the last one differed from the rest but had been swept to defeat by the general situation. At the urging of Raymond Ayoub, the Council of 27 August 1969 authorized the President to appoint a committee to investigate what opportunities to enter mathematics as a study or a career are available or are denied to various disadvantaged groups.

The Committee on Opportunities in Mathematics for Disadvantaged Groups consisted initially of Raymond Ayoub, chairman, Llayron L. Clarkson, Gloria Gilmer, Richard K. Lashof, Cathleen Morawetz, and David A. Sanchez.

In a report in January 1978, Raymond Ayoub as outgoing chairman summarized the work of the Committee. It had conducted a survey of graduate schools to determine the numbers of minorities in graduate programs, had proposed that a summer school for black mathematicians be conducted, and investigated the possibility of special programs. The second the third items did not materialize. The Committee on Women in Mathematics (see below) did appear. The Association for Women in Mathematics and the National Association of Mathematicians were established and took up some of the burden that might have fallen to the Committee.

The Committee became a joint committee with the Mathematical Association of America in 1986 and in 1988 became the Committee on Opportunities in Mathematics for Underrepresented Minorities.

COMMITTEE ON WOMEN IN MATHEMATICS

The Committee on Opportunities in Mathematics for Disadvantaged Groups presented a recommendation to the Council of 9 April 1971 that the Council "establish a separate committee... to deal with the question of women in mathematics." It was agreed that the problems of women as a "minority" appear to be different from those of racial groups. The Council established such a committee with the charge

> to identify and to recommend to the Council those actions which in their opinion the Society should take to alleviate some of the disadvantages that women mathematicians now experience and to document their recommendations and actions by presenting data.

The Committee on Women in Mathematics consisted of Mary W. Gray, Israel N. Herstein, Cathleen S. Morawetz, chairman, Charles B. Morrey, Jr., and Jane C. Scanlon.

The Committee reported to the Council of 24 January 1973 with six recommendations. Two were immediately adopted. The first was to maintain a roster of women with Ph.D.'s in mathematics. The second was to cooperate with the Mathematical Association of America to investigate the status of women mathematicians not in the Ph.D. track. Three more were amended and passed by the Council of 20 April 1973. One broadened the charge of the Committee on Academic Freedom, Tenure, and Employment Security to cover cases of discrimination on the basis of sex. Another encouraged the study of mathematics by women by presenting a more positive image of mathematics and the role of women in it. The last encouraged changes in current practices and policies in ways to equalize the employment opportunities of women in mathematics.

At the Council of 25 October 1974 the Committee was discharged. The Council approved a resolution that the roster be supplemented annually and updated every five years. The task was turned over to the newly formed joint AMS-MAA Committee on Women in Mathematics.

COMMITTEE ON COMMITTEES

The appointment of members of committees is a substantial part of the duties of a president. Members with diverse backgrounds and broad distribution in geography and types of employment are needed. These demands can exceed the span of knowledge of a president. To alleviate this problem President Lipman Bers stated his intention of establishing a Committee on Committees. His ad hoc committee made several recommendations. One was to establish an elected Nominating Committee and another to enlarge the list of duties delegated to the Executive Committee, both of which were done as has been noted elsewhere. Another was to establish a standing elected Committee on Committees, which the Council did in August 1975.

The Committee on Committees is appointed by the President for a term coincident with the President's own term. The Committee consists of seven members including the Secretary, another member not a member-at-large. It serves as consultant to the President in making appointments and to advise the President on the functioning of committees.

Fellowships

The Society has supported research fellowships. The AMS Research Fellowship was established in 1973 partly in response to the elimination of postdoctoral fellowships by the National Science Foundation. It was financed by contributions supplemented with general funds of the Society. It was agreed that the Society would contribute an amount equal to half the general contributions but at least $9,000 and at most $20,000. The amount of the fellowship was $10,000 with an expense allowance of $500. It was for research in mathematics strictly on the basis of merit, available to recent Ph.D.'s (or the equivalent, regardless of age). 'Recent' was interpreted to mean within the past five years. For 1976–1977 an additional requirement of citizenship or permanent residence in a country in North America was imposed.

Winners in this series of fellowships were the following:

1974–1975	Fred G. Abramson, James Li-Ming Wang
1975–1976	Terence J. Gaffney, Paul Nevai, George M. Reed
1976–1977	Fredric D. Ancel, Joseph A. Sgro
	A fellowship awarded to Eloise H. Carlton was declined.
1977–1978	David Vogan, Charles Patton, Duong-Hong Phong, Stephen Kalikow
1978–1979	Alan Danker, David Harbater, Howard Heller, Steven P. Kerckhoff, Robert C. McOwen
1979–1980	Scott W. Brown, Jeffrey E. Hoffstein, Jeffry N. Kahn, James E. McClure, Rick L. Smith, Mark Steinberger
1980–1981	Robert K. Lazarfield, Thomas H. Parker, Robert Sachs
1981–1982	Lawrence Man-Hou Ein, Mark Williams
1982–1983	Nicholas J. Kuhn
1983–1984	Russell David Lyons

The stipend for the fellowship had increased year by year until in 1983–1984 it was $24,000 plus an expense allowance of $1,000.

The NSF had restored its postdoctoral fellowships in fiscal year 1979, so that the Society fellowships and the NSF fellowships were reaching the same population. Accordingly the Society changed its fellowship to a mid-career fellowship, interpreted to mean five to ten years (four to ten in the first year) past the Ph.D. or equivalent. This was an area in which fewer other opportunities existed. These fellowships became available for the year 1984–1985. The stipend was $30,000 plus a $1,000 expense allowance. The citizenship or residence requirement was preserved. It was understood that the fellow could spread the fellowship over nine months plus one or two summers or could hold a teaching position with at most a half-time load and spread it over two years.

With the new eligibility requirement, the winners were as follows:

1984–1985	Richard Timothy Durrett
1985–1986	R. Michael Beals
1986–1987	Dinakar Ramakrishnan
1987–1988	Richard Hain, Bill Jacob
1988–1989	Steven R. Bell, Don M. Blasius, David Gabai

The number of fellowships had decreased in the interval 1982 to 1986 because of decreased contributions. At this point contributions were encouraged by a "negative checkoff" on the dues bill, with marked success.

The name was changed to Centennial Fellowship in 1987. The stipend was increased to $32,000 for 1989–1990.

Political and Social Questions

The Society has been involved in political or social issues that, according to some, fall outside the stated business of the Society of "the furtherance of the interests of mathematical scholarship and research." When these issues arise, there is a recurring difference of opinion whether their pursuit is proper business of the Society.

The situation was poorly defined prior to 1954 in that the matter of nonscientific questions was not considered at all in the bylaws. When Article IV, Section 8 was written the situation improved. The first paragraph of that section is as follows:

> The Council shall also have power to speak in the name of the Society with respect to matters affecting the status of mathematics or mathematicians, such as proposed or enacted federal or state legislation; conditions of employment in universities, colleges, or business, research or industrial organizations; regulations, policies, or acts of governmental agencies or instrumentalities; and other items which tend to affect the dignity and effective position of mathematics.

The remainder of the section is devoted to procedures. The section does not resolve the conflict in the minds of some, who continue to think that although the Council is now explicitly permitted to speak in certain situations it should nonetheless refrain.

The division of opinion was pointed up in the case of the loyalty oath at the University of California. It is not intended to rehearse the issues involved in this or other cases considered by the Society. Briefly, the Board of Regents of the University of California, acting under pressure, attempted to require a special oath of all employees. When this was protested, in 1950 an affirmation was added to contracts disclaiming membership in the Communist Party and other organizations as a condition of employment.

The Council of 1 September 1950 passed a resolution deploring this situation and asking the Regents to reconsider. The Council of 28 December 1950 passed a resolution saying that the Society would not meet at the University of California until the conditions previously condemned were alleviated. However, the Business Meeting of 29 December 1950 disapproved the action of the Council and instructed the Secretary not to release it beyond the membership. The disapproval concerned both substantive matters and Council authority. At this point the Council authorized a committee, which turned out to be M. H. Stone, chairman, T. H. Hildebrandt, and D. C. Spencer, to look into both the California situation and its implications with respect to Society policy. It is the latter aspect of the charge that is explored here. Stone and Spencer filed the report of the committee and Hildebrandt wrote a dissenting opinion for the Council of 4 September 1951.

Hildebrandt cited the declared purpose of the Society from its articles of incorporation, noted that the Society did not concern itself with conditions of employment of individual members, and stated that the proposed course of action meant taking over the functions of the American Association of University Professors or of a labor union. He further stated that if the Council wished to include questions about professional environment along with the stated duty to "formulate and administer the scientific policies of the Society" then there should be an amendment of the bylaws. The entire membership has no opportunity to approve or disapprove decisions of the Council. He went on to more substantive matters concerned with the current dispute, stating in particular that to bar meetings on the campuses of the University of California was childish and inquired in what way it furthers research. He recommended no further action.

The majority report examined both the situation in California and the general policy question, beginning with a historical summary of the state of affairs and the policy matters. It stated that the Society had never interpreted the statement in the Articles of Incorporation narrowly and cited education, the operation of selective service, the utilization of mathematicians in national security, and the maintenance of professional standards as examples. The Policy Committee was authorized to speak for the Society "on matters which concern the position of mathematics in such matters as proposed or enacted legislation concerned with science, problems concerning the effective use of mathematicians or potential members of our profession, and other questions which tend to affect the dignity and effective position of mathematics and related sciences, both nationally and internationally." (Similar wording was to be incorporated in Article IV, Section 5 of the bylaws.)

The report called for working to maintain appropriate conditions of professional activity and noted the value of the opportunity for free inquiry. The transfer of the costs of research to the federal budget points up the need for organized professional attention to the conditions of professional activity. The Society is the most natural and effective instrument.

The majority report interprets the bylaw authorizing the Council to "formulate and administer scientific policies of the Society" as lodging in the Council "the authority to decide when and how to act in carrying out the purposes for which the Society was incorporated." The report cited recent actions of the Council including the resolution of support for the National Science Foundation bill, which expressed concern over the loyalty oath in that bill, as examples of the interpretation being advanced.

The report supported the principle of representative government for the Society, spoke to the conditions that insure that the Council be representative, and delineated sound procedures, including recorded mail ballots of the Council on controversial or particularly important issues.

The report then went on to the specifics of the situation at the University of California, ending with a recommendation of a bar to meetings at the University of California for a three year period if conditions were not alleviated in the meantime. The report, with the exception of the section on procedures mentioned in the preceding paragraph, was approved.

This action of the Council did not settle the matter. In particular a committee to study questions of procedure in controversial matters and of Council membership was authorized by the Council of 27 December 1951. The latter issue rose over the fact that a number of Council members held their membership as editors and not as representatives.

The Council of 27 December 1951 passed a resolution directing the Secretary "to obtain, as a condition of holding a meeting, assurances that at any event scheduled in the program there will be no discrimination as to race, color, religion, or nationality, and that when accommodations and other facilities are provided these shall be provided to all attending the meeting." The issue arose concerning a prospective meeting at Auburn, AL, where the president of Alabama Polytechnic Institute had given appropriate assurances to Secretary J. R. Kline.

At about the same time as the loyalty oath was put in place by the Regents of the University of California, there was one ordered for all employees of the state of Oklahoma. On 1 October 1951 a joint committee with the Mathematical Association of America consisting of W. L. Duren, chairman, G. M. Ewing, and J. F. Randolph was authorized "to determine the facts concerning the loyalty oath at the Oklahoma Agricultural and Mechanical College and the University of Oklahoma and their effects on mathematics and mathematicians." In particular, there were dismissals or unplanned departures at Oklahoma A. and M. Again, it is not intended to review the facts and the arguments in the detailed and extensive report of the committee. The Council of 2 September 1952 issued a statement short of censure explaining briefly the unfortunate effects of a loyalty oath.

With this background the amendment now constituting Article IV, Section 8 was approved by the Council in 1953 and was presented in summary form to the Business Meeting in Kingston on 1 September 1953. In response to a motion by T. Radó, the Business Meeting requested that the full text of the amendment be circulated with a request for written comments from the membership. This was done in November, eliciting 32 letters, 18 expressing approval and 13 disapproval. Having viewed these letters the Council again recommended the amendment with the following statement:

> In presenting these amendments, The Council is not asking for broader powers, nor that it be made easier for the Council to speak with respect to matters which affect the dignity and effective position of mathematics.
>
> The Council believes, the world around us being what it is, that the Council will most probably have to consider such matters in the future.
>
> The Council therefore wishes an orderly procedure to be established to ensure that action will be taken only after careful study and only with due deliberation.
>
> The Council believes that these amendments guarantee such study and deliberation, and they are presented to you in this spirit.

A second amendment was presented at the same time, limiting the freedom of action of a Business Meeting. The last sentence of Article X, Section 1 had stated that:

> No matter of general business shall be considered at any meeting of the Society except the Annual Meeting, without the recommendation of the Council.

This was replaced by the current version, which reads:

> There shall be a business meeting of the Society at the Annual Meeting and at the Summer Meeting. A business meeting of the Society shall take final action only on business accepted by unanimous consent, or business notified to the full membership of the Society in the call for the meeting. Such notification shall be made only when so directed by a previous business meeting of the Society or by the Council.

Several issues have arisen subsequently that depended heavily on the use of these two amendments.

The Business Meeting on 25 January 1969 in New Orleans had unusually large attendance, estimated as being nearly one thousand late in the meeting.

Lee Lorch offered a motion requesting the Executive Committee to take steps to remove the meeting of the Western (now called Central) Section scheduled for 18-19 April in Chicago from that city. No reason was presented as part of the motion. However, it was the attribution to Mayor Richard Joseph Daley of repressive management of the Democratic National Convention recently held in Chicago that appeared to be in the minds of some of the supporters of the motion, which was passed.

The Executive Committee, which had earlier declined to move the meeting, did in fact move the meeting to Cincinnati, to the distress of some Society members who complained that politics was intruding into scientific activity. Six of the seven members of the Executive Committee were in attendance, with four votes in favor, one opposed, and one abstaining.

Parenthetically, one notes that there have been subsequent difficulties with Chicago. It may be the most central and readily accessible place in the Central Section for a meeting but in the turmoil over the Equal Rights Amendment the state of Illinois was one that did not ratify the amendment and so was out of favor as a meeting location until the time limit for ratification expired.

The Executive Committee was concerned how many of those in attendance and voting at the New Orleans Business Meeting were in fact members of the Society and considered restricting Business Meetings to members. This issue was resolved by the Council in April 1973, which adopted the following resolution:

> Each person who attends a Business Meeting of the Society shall be willing and able to identify himself as a member of the Society.

In explanation it was noted that:

> each person who is to vote at a meeting is thereby identifying himself as and claiming to be a member of the American Mathematical Society.

At the same New Orleans Business Meeting, Ed Dubinsky offered five resolutions with the following text:

> As a professional organization of academicians the members of the American Mathematical Society have the right and duty to take corporate action expressing their proper concern with conditions which affect the quality of civilized living and the evolving development of higher education. Specifically, the Society should adopt and support the following five resolutions which we respectfully propose for consideration at the next business meeting.
>
> 1. Resolved, that since scientific discovery by its nature requires complete open channels of information, it follows that classified

research is a contradiction in terms. Members should consider most seriously participation in any investigation under a contract restricting full exchange of information with learned men everywhere, and as a society we recommend that members seek to disengage themselves from such activity.

2. Resolved, that the American Mathematical Society urges each of its members to use his talents in ways that promote peace and to refrain from activities whose primary purpose is to promote warlike efforts.

3. Resolved, that a committee be appointed to study the causes and course of the current worldwide upheaval in relationships among faculty, students, and administration in higher education, with particular reference to the situation at San Francisco State College. This committee shall report to the members with recommendations for suitable action, in the *Notices* of the Society.

4. Resolved, that the *Notices* shall be open for letters and articles discussing issues which concern the members as scholars and citizens generally as well as mathematicians particularly.

5. Whereas the shortage of mathematicians in North American universities is different and greater among black and brown Americans than among whites, and whereas this situation is not improving, be it resolved that the AMS appoint a committee composed of black and third world mathematicians to study this problem and other problems concerning black and third world mathematicians, and report their conclusions and recommendations to the Society.

The Executive Committee and then the Council considered them in anticipation of their being formally offered to a Business Meeting. In the course of so doing, the Executive Committee formulated an additional resolution:

B. Whereas the American Mathematical Society encourages all persons interested in mathematical research to be members of the Society and whereas these members hold a wide variety of political and social views and have been welcomed to membership without regard to these views, resolved that the Society shall not attempt to speak with one voice for the membership on political and social issues not of direct professional concern and shall adhere closely to the purpose stated in its Articles of Incorporation of "furtherance of the interests of mathematical scholarship and research."

The Council voted to present resolution B to the membership by referendum with a favorable recommendation. The Council voted to present the five resolutions by Dubinsky by referendum with a statement that in view of its recommendation of resolution B it recommended a vote against each of the five. The referendum was conducted in May 1969. With about 7300 ballots, the votes were more than twelve to one in favor of resolution B and almost three to one opposed to resolutions 1, 2, 4, 5. The vote was more than five to one opposed to resolution 3. The secretary recalls being criticized for juxtaposition of the Council recommendations and the questions on the ballot.

In 1987, two resolutions were advanced by a group consisting of William P. Thurston, Michael Shub, Irwin Kra, Lipman Bers, Lee D. Mosher, Lucy J. Garnett, Linda Keen, and Jean E. Taylor. They were supported later by a petition with about 400 signatures. The text of the resolutions was as follows:

Motion 1. Many scientists consider SDI (commonly referred to as Star Wars) incapable of achieving its stated goals and dangerously destabilizing. Participation by universities and professional organizations lends a spurious scientific legitimacy to it. Therefore the AMS will lend no support to the Star Wars program. In particular, no one acting as a representative of the AMS shall participate in efforts to obtain funding for Star Wars research or to mediate between agencies granting Star Wars research money and those seeking to apply for it.

Motion 2. The AMS is concerned about the increasing militarization of support for mathematics research. There is a tendency to distribute this support through narrowly focussed (mission-oriented) programs which circumvent normal peer review procedures. This tendency, unless checked, may skew and ultimately injure mathematics in the United States. Therefore those representing the AMS are requested to direct their efforts towards increasing the fraction of non-military funding for mathematics research, as well as towards increasing total research support.

The Business Meeting of 22 January 1987 in San Antonio agreed to place these on the agenda of the Salt Lake City Business Meeting on 7 August 1987.

On the recommendation of the Committee on Science Policy (CSP), the Council of 25 April 1987 passed the following recommendations:

1. The Council instructs the Managing Editor of the *Notices* and the chairman of the *Notices* Editorial Committee to open its pages for comment related to two motions considered at the Business Meeting of 22 January, 1987.

2. The Council instructs the officers of the AMS to hold a mail ballot of the membership, after the January 1988 annual meeting

but before February 1988, on the substance of the two motions concerning issues of federal funding of research in mathematics.

In order to formulate the substance of the motions, the CSP offered recommendations that were modified by the Executive Committee and by the Council of 4 August 1987. This procedure resulted in five motions advanced by the Council as the text for the referendum. In setting the text, the Council neither approved nor disapproved any of the substance. This left the Society with two sets of resolutions, one for the Business Meeting of 7 August 1987 and the other for a referendum early in 1988. The Council requested that the motions at the Business Meeting be withdrawn in favor of the referendum and this was done.

The text of the five motions of the referendum was as follows:

Motion I. Many scientists consider SDI (commonly referred to as Star Wars) incapable of achieving its stated goals and dangerously destabilizing. Participation by universities and professional organizations lends a spurious scientific legitimacy to it. Therefore the AMS will lend no support to the Star Wars program. In particular, persons representing the AMS shall make no efforts to obtain funding for Star Wars research or to mediate between agencies granting Star Wars funds and people seeking these funds.

Note: SDI is an abbreviation for Strategic Defense Initiative.

Motion II. The AMS is concerned about the large proportion of military funding of mathematics research. There is a tendency to distribute this support through narrowly focussed (mission-oriented) programs and to circumvent peer review procedures. This situation may skew and ultimately injure mathematics in the United States. Therefore those representing the AMS are requested to direct their efforts towards increasing the fraction of non-military funding for mathematics research, as well as towards increasing total research support.

Motion III. Most seminal research in mathematics comes from individuals and small informal collaborations, not from large teams. The seriously low level of Federal funding for individual investigators documented in the 1984 David Report has recovered only slightly, and many of our best mathematicians are currently unable to find funding for research. Therefore, we urge that the Federal funding agencies not allow the recent trend toward large teams and big projects to compromise the strength through diversity of mathematics. We urge that in their continued attempts to bring mathematics funding into balance with that for related fields these agencies make every effort to increase the numbers of individual investigators to the levels recommended in the David Report.

Motion IV. Many advances in science and technolocy come from fundamental mathematics which has been developed without applications in mind. In recognition of the U.S. stake in the general excellence of mathematics, we urge agencies which fund research in mathematics to also fund a balanced proportion of basic research.

Motion V. We urge funding agencies in the mathematical sciences to solicit proposals openly and broadly and to assure that reviews of scientific merit are conducted by a diversified group of expert scientists.

The result at the close of the voting on 18 March 1988 was this:

MOTION	*I*	*II*	*III*	*IV*	*V*
Approve	4034	5193	6278	6355	6192
Do not approve	2293	1317	298	304	257
Abstain	719	539	469	385	590

The vote was evaluated on the basis of plurality. All five motions were approved.

With the result in hand, President G. D. Mostow made the following statement:

The first resolution, which was passed by 57% of those members who voted, reflects widespread skeptism in the mathematical community about the ability of the SDI program to achieve its stated objectives. It also reflects concern about SDI's incalculable cost and the waste incurred by premature deployment.

The second resolution, passed by 74% of those who voted, addresses the desirable norms for funding mathematical research. Resolution II also points out that the mission of the National Science Foundation is more closely matched to the aims of basic research than that of other agencies with more narrowly-focussed objectives.

The intent of the Society can best be read from the overwhelming approval of resolutions III, IV, and V which call for the support of diversity in mathematics through individuals and small groups, for the support of basic research by all agencies, and for reviews of scientific merit by competent scientists. It is clear from the last three resolutions that members of the Society seek support from all agencies that use mathematics.

Finances

The finances of the Society have become increasingly complex. It is not intended to give a historical analysis in any detail here. A few comments and a tabular summary is all that is attempted.

FUNDS

The Society used to keep its books on a basis of funds. In 1938 these included the Life Membership Fund, which was a reserve calculated actuarially to meet the obligations of the Society to life members. There were various operating funds, such as the Semicentennial Celebration Fund, the *Bulletin* Reprinting and Index Fund, the *Transactions* Reprinting and Index Fund, the *Transactions* Life Subscription Fund, and the Colloquium Publication Fund, in which funds accumulated in anticipation of future expenses. There were active operating funds such as the *Bulletin* and *Transactions* Funds, in which the income and expenses for each year were equal because there was an appropriation from general receipts if necessary to achieve that state. Salaries were apportioned among funds by estimating the fraction of time spent. There was no such item as overhead. It should be recalled that the kind of expense associated with overhead was minimal. There was no rent for space and management was by officers.

In 1938 there were also endowment and quasi-endowment funds, the former specifically designated as such by the Society or the donor, and the latter handled as endowment even though that status could be changed by the Trustees. These funds were the Bôcher Fund, the Cole Fund, the Moore Fund, and the beginnings of the Reilly Fund.

The Bôcher and Cole Funds supported the respective prizes. The Moore Fund was used at the discretion of the Council for publication of books or memoirs or for prizes. The Reilly Fund, opened in 1937, came from an estate in the process of being settled and was "to be used for the advancement of research in pure mathematics." It ultimately amounted to $23,600.

In the period since 1937, a number of bequests and special funds have appeared, a few of which are detailed here. The principal of the estate of

Robert Henderson came to the Society in 1961. Henderson was an actuary with the Equitable Life Assurance Society, the second Gibbs Lecturer (1924), and one of the Trustees at the time of incorporation in 1923. He served as trustee in 1924–1928 and 1931–1940. The bequest was placed in the general endowment and increased it by $547,223.20 from less than one hundred thousand dollars to more than six hundred thousand.

The Birkhoff Prize was endowed in 1967 with a gift of $2066 from the Birkhoff family, augmented with other gifts.

The Norbert Weiner Prize was endowed in 1967 with a gift of $2000 from the Department of Mathematics of the Massachusetts Institute of Technology.

The Steele Prizes were established in 1970 from a bequest of the residual estate of Leroy P. Steele, amounting to about $145,000, with some temporary items of income as well.

The Joseph Fels Ritt Memorial Fund was a bequest from Estelle F. Ritt, received in 1965 and amounting to $22,500. It was to be used for the publication of works in mathematics.

The status of funds was altered in 1986. Along with the Reilly and Ritt Funds the Society held several other quasi-endowment funds. These included the *Mathematical Reviews* Fund of $80,000 provided by the Carnegie Corporation when the journal was founded, royalties owed to Russian authors but not claimed in person and amounting to $69,452, the fund of $66,000 from the sale of the library, and the Friends of Mathematics Fund of $43,152 accumulated from smaller contributions. There was in addition a Dues and Publication Reserve accumulated since its establishment in 1964 to the amount of $148,671 and a Future Operations Fund of $356,041 that came from surplus of previous years. With the Cole Prize Fund of $2550, which was not a formal endowment, and some smaller funds the total was $796,417. All the above figures come from 31 December 1984. In May 1986, quasi-endowment funds were consolidated. The Dues and Publications and *Mathematical Reviews* Funds were added to the Future Operations Fund. The other quasi-endowment funds, excepting the Cole Prize Fund and the Russian Royalty Fund, were consolidated with the Friends of Mathematics Fund. The income from that fund is used for general purposes of the Society but is available for a special purpose should the Council and Board of Trustees designate one.

At the end of 1987 the total of endowment funds, including prize funds, stood at $2,018,200 and the total of quasi-endowment funds, including

Friends of Mathematics, Russian Royalties, and Future Operations, amounted to $3,971,411.

The Fiscal Year

The fiscal year of the Society in 1938 ran from 1 December to 30 November. That is, the Annual Meeting was the first event of the year.

In 1953 a firm of accountants was engaged. They made several recommendations, including change in accounting from a strict cash basis to a form of accrual basis. This was done effective 1 June 1953. Thus there is a short year, 1 December 1952 to 31 May 1953 in the tabulated figures elsewhere. The fisal year then ran from 1 June to 31 May.

The accountants also recommended the purchase of a National Cash Register accounting machine, which was done. This was the beginning of increasing mechanization of the accounting, which has progressed to a rather complete package on the mainframe computers in Providence.

The fiscal year changed again, becoming the calendar year for the year 1967. This corresponds again to a study of the system by a new Treasurer, W. T. Martin, whose term began in 1965. To accomplish the change there was a short fiscal year from 1 June 1966 to 31 December 1966.

Income, Expense, and Balance

The increasing bulk of the accounts of the Society is noted in the chapter on Planning and Organization. The handful of categories of income and expense tabulated year by year in [A] cannot realistically be continued here for the second fifty years because of the bulk that would materialize. Instead, members of the Fiscal Department in Providence have tabulated year by year the total receipts and the receipts in a few categories of interest and the total disbursements and the disbursements in principal categories. Further, they have provided summary balance sheets at five-year intervals with an analysis of the composition of the endowment and quasi-endowment funds.

Summary Balance Sheets
As of the End of the Fiscal Year
1937 to 1987 at Five Year Intervals

	1937	1942	1947	1952
ASSETS				
Cash and temporary investments	$12,176	$52,648	$86,875	$86,760
Publication inventories and work in process				
Property and equipment				
Other operating assets				4,667
Long-term investments	118,582	164,605	196,275	289,199
Total assets	$130,758	$217,253	$283,150	$380,626
LIABILITIES				
Deferred dues, subscriptions, and other revenues				
Other liabilities				
Total liabilities				
FUND BALANCES				
Operating funds	$38,216	$53,680	$91,875	$91,427
Quasi-endowment funds	12,595	95,333	107,094	179,793
Endowment funds	79,947	68,240	84,181	109,406
Total fund balances	130,758	217,253	283,150	380,626
Total liabilities and fund balances	$130,758	$217,253	$283,150	$380,626
ENDOWMENT FUND COMPOSITION				
General purpose funds	$74,629	$63,212	$74,101	$97,575
Prize funds	5,318	3,281	3,281	4,000
Pooled income fund				
Undistributed gains on investment transactions		1,747	6,799	7,831
Total	$79,947	$68,240	$84,181	$109,406
QUASI-ENDOWMENT FUND COMPOSITION				
Future operations				
Russian royalties				
Friends of Mathematics	$6,730	$24,521	$25,651	$27,552
Dues and publications	5,865	4,835	3,408	3,822
Library proceeds				50,000
Mathematical Reviews		60,000	65,000	80,000
Investment losses		3,537	4,386	5,550
Undistributed gains on investment transactions		2,440	8,649	12,869
Total	$12,595	$95,333	$107,094	$179,793

1957	1962	1967	1972	1977	1982	1987
$172,618	$510,922	$553,182	$722,674	$3,732,872	$1,559,073	$4,801,639
					580,463	1,560,959
	153,344	240,141	502,943	1,024,845	3,378,900	5,262,179
46,058	219,796	526,392	1,144,452	279,051	569,788	1,148,613
349,283	944,907	1,197,866	1,478,292	1,557,011	2,209,105	5,989,641
$567,959	$1,828,969	$2,517,581	$3,848,361	$6,593,779	$8,297,329	$18,763,031
$53,900	$180,472	$825,775	$1,593,440	$2,693,164	$4,003,349	$8,030,402
7,766	64,598	198,196	339,494	734,269	1,118,769	1,627,023
61,666	245,070	1,023,974	1,932,934	3,427,433	5,122,118	9,657,425
159,036	638,992	295,744	437,135	1,609,335	966,106	3,115,965
222,707	206,938	308,440	420,742	455,842	897,601	3,971,441
124,550	737,969	889,426	1,057,550	1,101,169	1,311,504	2,018,200
506,293	1,583,899	1,493,610	1,915,427	3,166,346	3,175,211	9,105,606
$567,959	$1,828,969	$2,517,581	$3,848,361	$6,593,779	$8,297,329	$18,763,031
$97,575	$650,798	$672,800	$673,319	$673,319	$673,319	$673,319
4,000	4,000	10,067	151,414	164,981	167,430	169,081
						5,000
22,975	83,171	206,559	232,817	262,869	470,755	1,170,800
$124,550	$737,969	$889,426	$1,057,550	$1,101,169	$1,311,504	$2,018,200
					$285,667	$3,747,911
			$81,717	$85,462	69,452	99,958
$27,551	$30,551	$30,551	39,552	39,874	60,135	123,572
2,524	1,514	54,707	68,400	87,777	126,641	
66,000	66,000	66,000	66,000	66,000	66,000	
80,000	80,000	80,000	80,000	80,000	80,000	
5,550	5,550	5,550				
41,082	23,323	71,632	85,073	96,729	209,706	
$222,707	$206,938	$308,440	$420,742	$455,842	$897,601	$3,971,441

RECEIPTS 1938–1987

YEAR	TOTAL RECEIPTS	DUES	JOURNALS	BOOKS
1938	35, 542	20, 986	7, 493	2, 109
1939	102, 730	21, 235	8, 367	2, 510
1940	57, 225	22, 201	15, 043	3, 121
1941	57, 958	22, 478	21, 401	3, 437
1942	49, 746	24, 082	14, 225	2, 872
1943	55, 278	23, 991	14, 049	4, 051
1944	65, 110	26, 519	21, 810	3, 924
1945	64, 189	26, 338	21, 010	3, 486
1946	75, 609	30, 076	24, 592	6, 451
1947	86, 440	30, 572	27, 723	8, 624
1948	128, 222	39, 413	36, 100	12, 734
1949	145, 918	40, 169	54, 056	17, 646
1950	262, 518	47, 846	64, 527	32, 045
1951	197, 716	60, 778	71, 051	25, 392
1952	216, 545	73, 218	92, 750	17, 049
1953	159, 906	50, 754	51, 916	7, 233
1954	293, 844	76, 620	92, 357	15, 681
1955	273, 025	78, 704	70, 319	13, 627
1956	290, 072	85, 175	70, 238	15, 666
1957	376, 914	95, 179	104, 724	20, 008
1958	520, 699	106, 317	119, 635	31, 975
1959	493, 011	119, 538	152, 062	31, 175
1960	707, 652	136, 323	188, 071	40, 389
1961	1, 013, 494	145, 826	217, 093	64, 865
1962	1, 293, 955	158, 947	316, 728	143, 378
1963	1, 357, 591	173, 549	394, 464	168, 222
1964	1, 660, 299	177, 816	333, 258	289, 628
1965	1, 440, 037	191, 697	493, 222	214, 329
1966	1, 780, 920	270, 497	746, 802	342, 851
1966	1, 037, 148	(9, 644)	128, 137	218, 605
1967	2, 385, 783	281, 835	808, 877	382, 611
1968	2, 614, 616	288, 277	1, 038, 565	479, 225
1969	2, 739, 575	317, 716	1, 366, 177	440, 725
1970	2, 879, 569	327, 893	1, 242, 873	453, 320
1971	2, 900, 743	347, 509	1, 454, 123	405, 037
1972	3, 000, 542	348, 550	1, 669, 915	419, 239
1973	3, 388, 927	420, 984	1, 883, 213	544, 896
1974	3, 691, 839	441, 808	2, 059, 752	592, 518
1975	4, 253, 208	445, 576	2, 433, 747	722, 701
1976	4, 534, 935	456, 335	2, 785, 595	665, 346
1977	4, 992, 731	477, 229	2, 964, 763	785, 499
1978	5, 368, 511	511, 716	3, 204, 527	770, 380
1979	6, 006, 018	675, 048	3, 372, 638	856, 101
1980	5, 935, 502	694, 210	3, 516, 037	559, 952
1981	6, 878, 740	725, 388	3, 786, 059	1, 169, 600
1982	7, 627, 085	749, 925	4, 558, 682	958, 104
1983	8, 186, 589	804, 816	5, 273, 876	654, 574
1984	9, 131, 910	891, 567	5, 993, 530	759, 517
1985	11, 247, 480	1, 045, 001	7, 285, 305	1, 003, 080
1986	13, 006, 984	1, 183, 552	7, 958, 448	1, 390, 366
1987	13, 429, 825	1, 284, 576	8, 456, 399	1, 235, 534

INVESTMENTS	MEETINGS	CONTRIBUTIONS	GRANTS	OTHER
3, 762		26		1, 166
4, 861		65, 019		738
5, 553	184	10, 036		1, 087
7, 284	238	2, 502		618
7, 362		510		695
11, 019		500	1, 250	418
11, 188		521		1, 148
12, 177		471		707
12, 589		471		1, 430
13, 662		3, 489		2, 370
14, 753		591	22, 560	2, 071
15, 508	14, 770			3, 769
16, 816	95, 155			6, 129
17, 241	937	358		21, 959
21, 880	4, 533	35		7, 080
11, 080		35	26, 617	12, 271*
22, 404	611		60, 414	25, 757
25, 219	355	3, 957	76, 335	4, 509
26, 486	903	5, 438	78, 555	7, 611
27, 409	556	1, 893	115, 654	11, 491
28, 007	725	6, 392	217, 934	9, 714
29, 455	1, 941	3, 725	141, 822	13, 293
32, 730	3, 050	12, 548	279, 960	14, 581
28, 326	5, 173	20, 462	465, 835	65, 914
39, 043		4, 698	526, 214	104, 947
46, 769		1, 748	460, 417	112, 422
43, 022		2, 099	665, 606	148, 870
39, 155	15, 520		346, 212	139, 902
43, 807	32, 335		188, 497	156, 131
28, 476	22, 900		508, 309	140, 365*
46, 398	58, 240		572, 223	235, 599
46, 146	56, 516	2, 717	486, 076	217, 094
67, 199	42, 603	4, 042	154, 041	347, 072
123, 472	54, 458	3, 855	260, 453	413, 245
93, 509	48, 298	3, 868	197, 978	350, 421
69, 740	53, 920	3, 536	109, 143	326, 499
80, 695	52, 739	5, 864	103, 303	297, 233
101, 951	40, 697	3, 918	139, 575	311, 620
151, 176	61, 602	5, 961	131, 594	300, 851
174, 641	63, 953	5, 480	57, 284	326, 301
204, 455	75, 000	5, 367	146, 257	334, 161
328, 073	68, 104	5, 772	121, 635	358, 304
466, 197	96, 405	7, 011	158, 588	374, 030
518, 234	137, 964	6, 412	154, 228	348, 465
490, 800	148, 689	7, 185	132, 047	418, 972
260, 344	151, 513	7, 058	328, 852	612, 607
185, 156	190, 573	22, 848	371, 328	683, 418
220, 614	132, 674	15, 276	424, 811	693, 921
343, 464	242, 446	11, 138	496, 174	820, 872
566, 114	241, 490	11, 137	430, 553	1, 225, 324
556, 780	308, 170	32, 961	441, 136	1, 114, 269

*short year

DISBURSEMENTS 1938-1987

YEAR	TOTAL DISBURSEMENTS	JOURNALS	BOOKS	MEETINGS
1938	31,506	19,399	2,352	
1939	43,186	25,149	5,998	
1940	49,006	35,400	2,727	
1941	47,429	32,974	2,970	
1942	48,996	32,980	6,254	
1943	48,017	32,014	3,756	
1944	45,482	32,604	1,922	
1945	45,477	35,097	1,530	
1946	67,414	50,254	5,720	
1947	82,803	60,154	6,474	
1948	123,536	73,686	28,577	
1949	144,918	86,836	28,998	2,166
1950	243,492	97,614	34,373	67,713
1951	189,242	106,362	24,490	3,960
1952	246,408	127,220	12,064	45,856
1953	111,757	68,022	2,130	4,135
1954	245,284	128,296	10,163	25,738
1955	272,420	140,343	10,110	27,752
1956	315,673	172,972	20,159	33,321
1957	391,113	197,235	22,795	38,678
1958	463,465	211,442	32,822	82,265
1959	486,707	253,969	29,841	48,685
1960	750,184	353,327	59,702	85,043
1961	947,200	439,311	88,428	118,376
1962	1,286,614	572,371	112,648	123,837
1963	1,343,602	595,975	269,173	129,438
1964	1,517,370	671,923	335,730	215,576
1965	1,396,734	782,908	256,445	106,275
1966	1,725,690	890,591	342,638	192,524
1966	1,206,790	632,276	223,335	89,129
1967	2,185,518	1,078,405	357,632	225,932
1968	2,659,247	1,208,970	404,570	183,407
1969	2,845,292	1,586,780	398,499	127,564
1970	3,058,813	1,588,330	441,906	186,763
1971	3,048,121	1,871,457	485,187	126,028
1972	3,445,881	1,930,960	435,345	161,421
1973	4,098,381	2,214,534	534,712	182,022
1974	4,252,190	2,539,888	520,710	216,837
1975	3,955,565	2,685,792	360,998	236,373
1976	4,095,207	2,834,821	424,130	243,955
1977	4,392,286	2,919,004	496,953	239,455
1978	5,571,795	3,524,995	409,223	284,452
1979	5,948,318	3,962,479	638,637	344,430
1980	6,932,806	4,624,416	412,551	393,695
1981	9,280,566	4,814,906	1,429,964	375,490
1982	8,909,104	5,057,299	811,696	606,214
1983	9,909,278	6,042,632	972,026	770,524
1984	10,302,047	6,359,499	784,712	694,098
1985	10,988,233	6,035,128	1,229,444	827,609
1986	11,411,395	6,807,766	1,048,086	624,619
1987	12,995,497	7,907,351	877,737	898,972

MEMBERSHIP ACTIVITIES	PROPERTY PLANT EQUIPMENT	OTHER
	630	9,125
	824	11,215
	31	10,848
	52	11,433
	44	9,718
		12,247
		10,956
		8,850
		11,440
	769	15,406
	1,050	20,223
	590	26,328
	501	43,291
	2,199	52,231
	759	60,509
	311	37,159 *
	4,389	76,698
		94,215
	5,540	83,681
	8,615	123,790
	11,510	125,426
	10,122	144,090
	89,295	162,817
	41,339	259,746
	114,253	363,505
	35,223	313,793
	19,193	274,948
	16,033	235,073
	19,964	279,973
	14,481	247,569 *
	151,415	372,134
	108,917	753,383
	46,816	685,633
29,971	32,185	779,658
37,516	26,075	501,858
44,583	314,083	559,489
29,735	470,237	667,141
47,785	488,326	438,644
45,553	72,055	554,794
59,299	32,373	500,629
61,100	88,069	587,705
51,151	675,127	626,847
54,036	412,434	536,302
66,154	941,145	494,845
70,889	1,564,019	1,025,298
108,359	531,212	1,794,324
63,000	666,163	1,394,933
314,153	1,276,541	873,044
454,773	1,440,929	1,000,350
508,899	448,297	1,973,728
510,107	1,237,307	1,564,023

* short year

The Headquarters

During its first fifty years and beyond, the business and editorial office of the Society was at Columbia University in one of several locations.

THE MOVE FROM NEW YORK TO PROVIDENCE

The move of the office of the Society from New York to Providence was connected with three problems of the Society, office space, library, and finances.

The Society library had begun through journal exchanges. There was sufficient activity that the office of Librarian was established in 1892. Journal exchanges were a major source of library material. It was a reference library and its volumes were available on loan to members of the Society. Publishers of books up until World War II regularly provided two copies of books, of which only one was needed for review. A catalog of the library appeared regularly. See [A] for more detail.

The library was housed in Low Library of Columbia University on terms quite favorable to the Society. By 1950 it had grown to about thirteen thousand items, of which two thirds were bound periodicals. Its value was estimated to be at least $50,000. The offices of the Society, where the work included membership records and the handling of manuscripts for publication by a staff of four, were also in Low Library. Both the Society and the University felt the shortage of space that the arrangement engendered as each grew. The employment of the first executive director in October 1949 created an additional demand for space.

The Society was short of funds. A Committee on Additional Revenue, under the chairmanship of J. L. Walsh, reported to the Council of 26 December 1946. Among its recommendations was the suggestion "that the Council study the possibility of selling the library, or a substantial part thereof, to obtain capital and reduce clerical work." A motion to that effect failed. This was not the first time that the Society had declined to sell the library. It was stated that a year or two earlier it had declined to sell it to Columbia University.

To solve this complex of problems, the thought arose of bargaining with Columbia over the library on one hand and the office space on the other. A house on West 117 Street owned by the University was a piece of space under consideration. The Trustees had already authorized a committee to study the problems and the Council took the position that it should be a joint committee of the Council and the Trustees. It consisted of T. H. Hildebrandt, chairman, Arnold Dresden, J. R. Kline, Marston Morse, P. A. Smith, and Warren Weaver.

Negotiations with Columbia proceeded. At the same time, the Council of 28 October 1950, subject to the approval of the Trustees, authorized the executive director to solicit bids for the library. Should bids in excess of $57,000 be received, the Executive Committee was authorized to choose for the Council among them, giving due regard both to the bid and to the prospect of effective use of the library. Otherwise, the Executive Committee was to report to the Council. The Trustees of 21 November 1950 voted however to postpone action on the Council recommendations, though it appeared that Columbia might not be interested in the purchase of the library and might well not have sufficient space available to meet the needs of the Society.

The Council went on record in favor of the purchase of a building, not necessarily in New York, adequate for all the offices of the Society.

At the meeting of 13 January 1951 the Trustees agreed to consider the offer of assistance of Columbia in finding space near the university and to explore further the opportunity of establishing the offices of the Society at or near Brown or Yale University, these two proposals being more favorable than others that had been presented. The possibility of moving west of New York was also to be studied.

The Trustees authorized the executive director to prepare a prospectus for the sale of the library. The Executive Committee considered the five bids received in response. The Committee recommended to the Council that the Society accept the bid of $66,000 from the University of Georgia, this bid both being the highest and meeting the requirements of the Council. Columbia had bid on only parts of the library. The Council accepted the recommendation of the Executive Committee and the Board approved. In giving its approval, the Council asked the Librarian to consider the policy of continuing exchanges with other societies, as for the benefit of *Mathematical Reviews*.

At the same time the Executive Committee recommended to the Council that the headquarters of the Society be established in Providence in a building known as the Delta Upsilon House at 80 Waterman Street to be leased from Brown University. The Council and then the Trustees approved. The move took place in the week of 17 September 1951.

The first $50,000 received from the sale of the library was put into endowment. The remaining $16,000 was made available for the expenses of transfer of offices to Providence, with the understanding that over a period of five years equivalent funds were to be added to the Library Proceeds Fund.

THE RENTED LOCATIONS

The Society headquarters stayed at 80 Waterman Street from 1951 to 1956. During 1955 the Society considered a property at 190 Hope street which was in the estate of Miss Elizabeth M. Chace. The will specified that if the property was sold then Brown University was to have the opportunity to buy it. The asking price was $50,000. An agreement was reached whereby Brown University purchased the property. The sale price appears to have been $45,000. The Society was then to purchase the property from Brown and authorized the purchase. In the meantime the Society leased the property and moved in. While owned by Brown, the property was exempt from local taxes. The legislature passed a tax exemption bill in favor of the Society but the governor vetoed it. The Society continued to lease from Brown until a tax exemption of $75,000 was obtained in April 1959. Then the Society purchased the property for $45,000. After 1962 it was used briefly for storage though both inefficient and a fire hazard. The property was sold to Brown in 1965 for $50,000.

From 1962 to 1968 the Society was housed in part of Butler Hospital. This was a hospital for the mentally ill. There were entrances and communicating corridors used by both the Society and the hospital, leading to a continuing joke about distinguishing the Society staff from the patients.

From 1951 to 1965, the *Mathematical Reviews* was housed with the headquarters. The later locations of *Mathematical Reviews* have been detailed with the discussion of that journal.

For several years the Trustees had been thinking about building and owning a headquarters building. In 1968 the Society moved to a new building on South Main Street. This was the Heritage Building, where the Society leased two of the four floors and had an option to buy the building. When it came time to exercise the option, at a price of about $1,700,000 there were several factors to consider. The prospect of being a landlord did not appeal and the possibility of securing a tenant seeming none too sure. The cost seemed high. It was not clear that the neighborhood would be developed in a desirable manner. Consequently the Society decided to build at a cost of about $800,000.

It should be noted in retrospect that the Society has added to the building it built to the point that it would have filled the Heritage Building with its own operations, that the area of South Main Street has flowered, and that the market value of the Heritage Building has multiplied. That is, in hindsight it would have been a good buy.

201 Charles Street

The Society built a one story building of about 22,000 square feet at 201 Charles Street in an area of redevelopment in Providence and occupied it on 15 May 1974. In an effort to keep the cost down, the structure was not planned to allow for a second story, a decision that one has come to regret. Subsequent enlargements have included two wings totalling 2700 square feet completed in 1978. The interior has been repeatedly remodeled through the use of modular cubicles to accommodate people more efficiently and assure freedom from distraction in working conditions.

The Society has a tax exemption of $2,000,000 through 1990.

For a number of years the Society had rented warehouse space for storage of back numbers. From April to October 1981 a two-story warehouse was built for that purpose adjoining the headquarters and connected to it. Office space in the main building has subsequently been expanded by moving certain operations such as printing from the main building to the warehouse.

The Staff

The staff of the Society prior to World War II consisted of four persons. There were no titles listed in the budget but the duties were roughly office manager, secretary, accountant, and editorial assistant. However, it was an office in which everyone had to do everything. In particular, these same people together with volunteers were the staff who handled meetings.

Beginning in 1940 there was a staff of three for *Mathematical Reviews,* increasing to four in 1946 and five in 1948.

By 1952, when the general headquarters and *Mathematical Reviews* were all in Providence at 80 Waterman Street, the total staff had increased to 19.

In 1955 with the advent of J. R. Curtiss as executive director, a formal table of organization appears in the minutes of the Trustees. It shows the executive editor and a director of Special Projects reporting to the executive director. There are departments of administrative services, fiscal services, editorial services, and a mathematical sciences service bureau. There was an office manager who supervised departments, planned office procedures and correspondence control, and was in charge of personnel and stenographic services. Administrative Services covered membership services such as membership records, arrangements for meetings, technical editing of the *Notices,* and Trustees' business and office services such as purchasing, printing contracts, central files, and plant management. The Fiscal Department had an accounting section concerned with bookkeeping, billing, disbursements, and financial reports, and a sales section that dealt with orders and inventory. The Editorial Department took care of the technical editing of scientific publications and supervised printing and publishing of books and journals. The Mathematical Sciences Service Bureau was concerned with administrative and publication projects performed jointly with other mathematical organizations. The staff of 1954-1955 consisted of 24 persons, three in the executive director's office, three in accounting, three in membership, one secretary, four in sales, two in editorial, one on building maintenance, and seven on *Mathematical Reviews*. In addition there were some part-time employees.

When the Society moved in 1962 the number of employees was up to about 82, including *Mathematical Reviews*. When the move was made to the Heritage Building in 1968 there were about 87 employees in Providence and 26 more with *Mathematical Reviews* in Ann Arbor.

In 1973 when the Society occupied its own building, the number of employees in Providence was up to 116 and the number in Ann Arbor was up to 35.

The number of employees budgeted for Providence in 1988 was almost 156 full-time equivalents. It should be noted that more than four were allocated to sale of services, such as composition and printing for others at a price. The budgeted number of employees in Ann Arbor was about 75.

In counting personnel, it was not always possible to distinguish between numbers of actual employees and budgeted positions, though the two numbers were always close, the latter slightly larger.

Archives

The Society has considered from time to time the establishment of formal archives. Initially, papers were in the hands of the secretary. From early days, the only documents still in existence are minute books of the Society and of the Council, some programs of meetings, and some lists of members and speakers. The minutes of the Society come to an end in 1912. There is a gap in minutes of the Council from 1907 to 1913 because these were destroyed in a fire. Files of correspondence of successive secretaries beginning with R. G. D. Richardson in 1920 are in the possession of the secretary and await deposit in the archives at Brown University. The writer does not know whether the correspondence files of Secretaries Thomas Scott Fiske and Frank Nelson Cole still exist.

An effort was begun in the Providence office in the 1960s to organize the files of the secretaries but it came to a halt partly as an economy move when there was a slight reduction in staff. In 1975, R. L. Wilder raised with the secretary the question whether the Society has formal archives, as he was seeking a place for his papers and Lucille Whyburn was doing so for the papers of G. T. Whyburn. The University of Texas was a natural place that they had in mind because of association with R. L. Moore.

In December 1976, following exploration and correspondence with many people by H. L. Alder, President Elect of the MAA, a joint AMS-MAA Fact-Finding Committee on Archives was established. It consisted of Alder as chairman, Judith V. Grabiner, Robert B. Greenwood, Everett Pitcher, David P. Roselle, and Wilder. The intent was possibly jointly to select a site for archives of both organizations.

Wilder and then Alder had examined the facilities of the Humanities Research Center (HRC) at the University of Texas, which professed interest and which held the papers of R. L. Moore, President of the Society in 1937–1938.

The desirability of archives separate from and not in complete control of the Society posed problems of access insofar as these are needed at irregular intervals for the conduct of Society business. On the other hand, order could be introduced into the files, as by indexing and cataloging, and they could be made more readily available to interested parties in a public place.

During 1977 a contract with HRC was formulated and was examined both by principals and attorneys. With some revision the MAA found it satisfactory, the Board of Governors approved it in January 1978, and it was signed and became effective for the MAA soon thereafter.

The Trustees were not satisfied with the proposal to place the archives at HRC and at their meeting of 10–11 December 1977 decided to appoint a committee, consisting of Cathleen Morawetz, F. P. Peterson, and Calvin C. Moore, to explore further. In February 1978 Alex Rosenberg was co-opted. The matter was discussed at length by the Trustees in May of 1978. The committee initially provisionally favored the HRC proposal. It developed that Brown University was quite interested in receiving the Society archives. Detailed examination of draft contracts showed that Brown appeared to offer a more favorable contract both financially and in terms of available services.

The Trustees wished still more information concerning the Brown contract and the matter was deferred to December 1978. At that meeting the committee of Trustees recommended that the Society deposit its archives at Brown and the Executive Committee and Board of Trustees approved. This was not the end of the matter. A question of jurisdiction was raised in the Council of 23 January 1979. It was the position of the Trustees "that the establishment of archives is a matter of care or disposal of property of the society and that the Trustees, acting as usual with the advice of the Executive Committee, have full jurisdiction." However, the joint Fact Finding Committee had not been discharged. It took the position in a report from Alder that the Society was about to act in a manner contrary to the recommendation of the Fact Finding Committee and that "much more than the care and disposition of property is involved in deciding on a location for the Society's archives." Moreover, it proposed to prepare for the Council a statement showing among other things the tentative nature of the Brown proposal and the desirability of having AMS and MAA archives together at HRC. R H Bing also asked that the Council consider the issues. The Council did agree to receive the proposed report of the Fact Finding Committee. The committee suggested a presentation in August 1979. However, the secretary was instructed to schedule it for 20 April 1979 and to invite Alder, the chairman of the committee, to appear at the Council meeting. This produced a second temporizing statement from Alder for the Fact Finding Committee. After extended discussion, the Council of 20 April 1979 voted to wait until August 1979 to hear the report.

The Executive Committee and Board of Trustees of 11–13 May 1979 voted to supply the Fact Finding Committee with all available documents but noted that the Trustees had already taken a position.

At its meeting of 22 August 1979 the Council received a majority report from the Fact Finding Committee, a report from a minority of one, and background information, the total coming to 152 pages. The Council considered the first recommendation of the committee that the AMS "consider postponing the final decision [on its archives] once again, this time to seek, on its own, the evaluation of independent experts, both archivists and historians." This was replaced by a substitute motion "that the Council requests the Board of Trustees to rescind the action taken in December 1978 [concerning archives] and that any further action be laid over indefinitely until such time as the Council or Board of Trustees chooses to put it on the table." The Fact Finding Committee concurred with this position and the Council passed the motion.

In view of this action of the Council, the Board of Trustees of 9–11 November 1979 "instruct[ed] the executive director to defer the accomplishment of a contract with Brown University for the establishment there of the Archives of the Society."

The Society understands that the interests of the HRC changed with a change in personnel and that the Mathematical Association in fact deposited its archives with the library of the University of Texas in Austin rather than with the HRC.

It may be just as well that the Society postponed action on a contract with Brown in that Brown was rebuilding some of its library structures and may not have been quite ready to proceed. In any event, conversations between the Society and Brown were resumed in 1984 and resulted in the signing of a contract in January 1986.

The contract provides several features that the Society values beyond the care for archives. One is advice in the establishment of a Record Management Program for the Society that will assist in separating non-current from current records and in establishing retention schedules and disposition of non-current records, including those of archival value. Another is the storage of a single copy of every item published by the Society for the purpose of future reproduction by the Society.

In order to proceed, grant money is to be sought. The National Historical Publications and Records Commission in late 1987 provided a grant of $4668 for a records management planning project. A consultant on records management, Victoria A. Davis, undertook early in 1988 to provide the required advice. With this in hand the Society and Brown University are in a position to apply for grant money to establish the archives, effect the initial transfer of material, and process it.

Index